Climate Finance

A project of the Institute for International Law and Justice (IILJ)
at New York University School of Law

Climate Finance

Regulatory and Funding Strategies for Climate Change and Global Development

EDITED BY

Richard B. Stewart, Benedict Kingsbury, and Bryce Rudyk

A publication of the New York University Abu Dhabi Institute

New York University Press

NEW YORK AND LONDON

NEW YORK UNIVERSITY PRESS
New York and London
www.nyupress.org

A Cataloging-in-Publication Data record for this book
is available from the Library of Congress.

ISBN-13: 978-08147-4138-2 (pbk.)
ISBN: 0-8147-4138-X

New York University Press books are printed on acid-free paper,
and their binding materials are chosen for strength and durability.
We strive to use environmentally responsible suppliers and materials
to the greatest extent possible in publishing our books.

Manufactured in the United States of America

p 10 9 8 7 6 5 4 3 2 1

Contents

PART II Proposals for Climate Finance: Regulatory and Market Mechanisms and Incentives

Acknowledgments

This book is a result of the Climate Finance: Financing Green Development project of the Institute for International Law and Justice (IILJ) at New York University School of Law, undertaken jointly with NYU's Frank Guarini Center on Environmental and Land Use Law (CELUL). The project has received very generous support and encouragement from the NYU Abu Dhabi Institute, an innovative partnership of NYU and Abu Dhabi. Our colleague and friend NYU President John Sexton was a driving force, together with Abu Dhabi and NYU leaders, in the inception of this partnership and its embrace of this project. Mariët Westermann, the Provost of NYU Abu Dhabi, and Philip Kennedy, the Faculty Director at the NYU Abu Dhabi Institute, made possible the project's major conference on climate finance in Abu Dhabi in May 2009, in which developing and developed country participants from governments, the climate finance industry, multinational businesses, international organizations, NGOs, and academic and research institutions presented and discussed preliminary versions of the proposals and analyses that appear in the following chapters.

We are grateful to all authors and conference participants, to NYU Abu Dhabi for their support and for funding much of the work needed to produce this book, to the Emirates Center for Strategic Studies and Research for the kind hospitality in hosting the conference, and to NYU Law School, Dean Richard Revesz, and the Hauser Global Law School Program. Our faculty colleagues Kevin Davis (Director of the closely linked Financing Development Project), Robert Howse (on trade and investment issues), and Lily Batchelder and Mitchell Kane (on tax issues) contributed much to the planning of this project, as well as their own essays. We acknowledge with thanks the support for faculty research provided by the D'Agostino and Greenberg Fund and a grant to the IILJ from Carnegie Corporation of New York for work on global administrative law and specific issues concerning developing countries, to which Simon Chesterman and Angelina Fisher contribute greatly. We also thank Tom Heller, Ngaire Woods, and Bernard Heikel for invaluable support and wise advice, and Marcel Brinkman and Henry Derwent for serving as panel chairs.

Toni Moyes, formerly Research Fellow at NYU's CELUL and now with the New Zealand Ministry for the Environment, played a key role in the formulation of the project. As we sought to construct a unified field of climate finance from a very diffuse body of practice and ideas, her imagination and persistence in synthesizing the vast literature and identifying participants and perspectives were invaluable. She has been succeeded in this role by Bryce Rudyk, Research Fellow at CELUL, to whom the senior editors also express deep appreciation. Bringing together the participants at the conference in Abu Dhabi was a team effort. At NYU Law School, Sarah Dadush, Sun-Young Suh, Alma Fuentes, Basilio Valdehuesa, and Meera de Mel, as leader of the team, worked indefatigably to plan and organize the conference and also the climate finance website (at www.iilj.org) which accompanies this book. Sarah and Sun-Young as rapporteurs also furnished extensive notes and summaries of the conference, which helped greatly in the drafting of parts of this book. Considerable assistance was provided both in New York and Abu Dhabi by the staff of NYU Abu Dhabi: Maura McGurk, Larry Fabian, Brett Heger, Jennifer Sloan, Nils Lewis, Antoine El Khayat, Diana Chester, Peter Christensen, Catherine Kosseau, Brooke Beyer, Carol Gardner, Dean Williamson, and Hilary Ballon.

This is an unusual volume in condensing into thirty-six short and trenchant policy essays considerable thought, analysis, and research. Steve Maikowski, the director of NYU Press, has provided exceptional leadership and guidance in taking on this project and enabling us to produce and distribute this book in very timely fashion. We also thank the other staff at the Press. At NYU Law School, James Chapman, John Wunderlin, and Rachel Jones did sterling work in editing, checking, polishing, and proofreading the essays under considerable time pressure.

Finally, we are deeply thankful for the support of our much-loved families and friends.

Richard Stewart, Benedict Kingsbury, Bryce Rudyk
New York City, September 2009

Foreword

NYU Abu Dhabi and the Sustainable Environment

Mariët Westermann
Provost, NYU Abu Dhabi

Philip Kennedy
Faculty Director, NYU Abu Dhabi Institute

This book is the first volume of policy papers issuing from the NYU Abu Dhabi Institute. It demonstrates NYU Abu Dhabi's commitment to scholarship on matters that have critical significance in the world today. And we consider it fitting, with the selection of Abu Dhabi as home for the International Renewable Energy Agency and the ground-breaking work of Masdar, that this volume tackles one of the most pressing global issues: climate change.

Climate change mitigation and adaptation requires complex thinking across a wide range of fields: science, economics, finance, public policy, and law. Effective action demands the collaboration of institutions on all continents. Resulting from a conference on novel mechanisms and frameworks for financing climate change abatement—organized jointly by NYU's School of Law and the NYU Abu Dhabi Institute—the papers in this book exemplify this cross-disciplinary and inter-continental approach. Abu Dhabi is an apt location for such a meeting of minds and call to action. As a flat, low-lying, desert country with an extensive Gulf coastline, at a latitude of 24° N, Abu Dhabi has a particularly direct understanding of the risks attached to even a small rise in global temperatures and sea levels. The Emirate's long-term strategy to diversify its economy beyond hydrocarbon resources includes a major commitment to renewable energy research and the policy work required to translate such new knowledge into economically viable applications.

Just as Abu Dhabi is taking a leadership role in fostering research and meaningful dialogue toward a sustainable environment, NYU Abu Dhabi is developing its new campus on Saadiyat, a natural island of great beauty, in a manner that will minimize its environmental impact. In the same spirit, NYU Abu Dhabi's undergraduate curriculum offers a multidisciplinary concentration on the environment. Students will study and research environmental science to understand the scientific foundations of climate change, responsible use of natural resources, and sustainable development. They will learn about a wide range of strategies that can inhibit damaging and irreversible environmental change, while making connections between abstract scientific concepts, the physical world around us, and local and global policy.

As a full partnership between Abu Dhabi and NYU's global community of scholars and students, NYU Abu Dhabi fosters the development of well-founded and practical solutions to the most challenging problems of our century. We hope that this book is but a first step.

Abu Dhabi, September 2009

Summary of Key Findings and Recommendations

Meeting the imperative of achieving major reduction in greenhouse gas emissions in developing as well as developed countries, without sacrificing urgently needed development, requires far greater attention to the emerging subject of climate finance than it has yet received. To achieve the necessary mitigation of climate change in developing countries, *additional* investments of €55–80 billion each year during the period 2010–2020 may be required, rising to USD 92–96 billion per year by 2030. Carbon markets are part, but only part, of the solution. Innovative financing, regulation, and governance are essential. The following strategies are proposed:

- A variety of new arrangements to generate public and private climate finance and engage developing countries in mitigation are needed; a single uniform design is neither feasible nor desirable. Ideally, they should be designed to support and not retard the future adoption by major developing countries of emissions caps.
- A suite of revised or new market-based mechanisms must be developed to mobilize very large increases in private investment in developing country mitigation. These include a reformed Clean Development Mechanism (CDM) and credit offset trading systems established pursuant to domestic cap-and-trade climate regulation by developed countries.
- These mechanisms must leverage private investment in order to achieve net climate benefits and secure long-term low carbon development.
- Carbon markets cannot be autonomous; they must be structured, regulated with developing as well as developed country involvement in their design and governance. Governance arrangements should be transparent and provide for appropriate mechanisms for account-

ability to non-state actors, including investors and non-governmental organizations.

- Linkages among national and regional regulatory/trading systems through allowance trading and transfers of offset credits will play a key role; achieving them will require coordination among governments.
- Governance arrangements and the determination of conditions on official development assistance (ODA) must be changed significantly to enhance developing countries' roles, build trust, and assure climate-sustainable development. Greater integration or coordination of international ODA mechanisms is also needed.
- The new arrangements for both private investment and ODA must be structured to match with the different types and costs of mitigation opportunities available in developing countries.
- New institutional arrangements are needed to recognize, facilitate, and coordinate the diversity of decentralized climate initiatives among both developing and developed countries.
- World Trade Organization (WTO) trade rules need to be interpreted and applied to accommodate domestic climate-related regulatory measures, including border carbon adjustments to deal with competitiveness and leakage issues and mitigation technology subsidies, so long as they are non-discriminatory and not protectionist.
- The WTO and developing countries need to develop additional capacities to monitor and respond to adoption of climate-related domestic measures that impact trade in potentially distortionary or protectionist ways.
- Changes in tax laws, including a degree of harmonization among national tax systems, are needed in order to avoid creating market distortions and regulatory inefficiencies in trading-based climate regulatory systems.

About the Contributors

Lily Batchelder is Professor at NYU School of Law. Her research centers on income taxation, wealth transfer taxation, tax incentives, and social insurance. She previously worked at Skadden, Arps, Slate, Meagher & Flom, where she focused on transactional and tax policy matters.

Eric C. Bettelheim is a Founder and Executive Chairman of Sustainable Forestry Management Ltd. His publications include coeditorship of the Royal Society's volume on free-market approaches to land-use change and forestry, which includes an article of his on carbon sinks.

Daniel Bodansky is the Emily and Ernest Woodruff Chair in International Law at the University of Georgia, and is the US-nominated arbitrator under the Antarctic Environment Protocol. Formerly he served as US State Department Climate Change Coordinator and advised the UN in climate change and tobacco control.

Marcel Brinkman is Associate Principal in McKinsey & Company's London Office and a member of McKinsey's Corporate Finance practice. He co-leads the McKinsey's Environmental Finance service line and leads the Project Catalyst Carbon Finance work.

James Chapman is Research Fellow at the Center for Environmental and Land Use Law at NYU School of Law. His research focuses on establishing links between distinct emissions trading schemes and on nuclear waste law and policy and its implications for low-carbon development.

Rae Kwon Chung is Ambassador for Climate Change for the Republic of Korea. He received the 2007 Nobel Peace Prize for his work with the IPCC and was Director for the Environment and Sustainable Development Division at the UN Economic and Social Commission for Asia and the Pacific.

Sarah Dadush is Fellow at the Institute for International Law and Justice at NYU School of Law. She works primarily on the Institute's Financing

Development program. Her research focuses on the regulation of immigrant remittances and innovations for development financing.

Kevin E. Davis is the Beller Family Professor of Business Law at NYU School of Law. His current research centers on contract law, the governance of financial transactions involving developing countries, and the general relationship between law and economic development.

Henry Derwent is President and CEO of the International Emissions Trading Association. Previously he was the international climate change director for the UK government and served as the prime minister's special representative during the UK G8 presidency in 2005.

Navroz K. Dubash is Senior Fellow at the Centre for Policy Research in New Delhi and Associate Professor at the Centre for the Study of Law and Governance, Jawaharlal Nehru University, Delhi. He has held positions at the National Institute of Public Finance and Policy (Delhi), the World Resources Institute, and the Climate Action Network.

Antonia Eliason is an Associate in Allen & Overy's London office. She clerked at the European Court of Justice and at the Legal Affairs division of the WTO. She was also a visiting research fellow at the World Trade Institute, where she researched climate change mitigation and issues related to the SPS Agreement.

Mark Fulton is global head of strategic planning at Deutsche Asset Management Climate Change Advisors in New York. He has extensive experience in research and management in private banking.

Arunabha Ghosh is Oxford-Princeton Global Leaders Fellow. He previously worked as Policy Specialist at UNDP's Human Development Report Office in New York, where he authored the 2006 HDR and coauthored the 2005 and 2004 editions.

Luis Gomez-Echeverri is with the Global Energy Assessment Program at the International Institute for Applied Systems Analysis. He has held positions at UNDP and the Secretariat of the UNFCCC and led the UN secretary-general's WEHAB (Water, Energy, Health, Agriculture, and Biodiversity) initiative.

Thomas Heller is Professor at Stanford Law School and Senior Fellow at Stanford's Freeman Spogli Institute for International Studies and at the Woods Institute for the Environment. He is Executive Director of the Climate Policy Initiative, Climate Advisor to George Soros, a member of the core team of Project Catalyst, and advises Working Group III of the IPCC.

Robert Howse is Lloyd C. Nelson Professor of International Law at NYU School of Law. He is a contributor to the American Law Institute project on WTO Law, has been involved in several NAFTA arbitrations, and is a core team member of the Renewable Energy and International Law project.

Mitchell A. Kane is Professor of tax law at NYU School of Law. His current research focuses on the intersections between tax law and policy and economic development, climate change and corporate governance. He clerked for the Honorable Karen LeCraft Henderson of the US Court of Appeals for the D.C. Circuit and was a tax associate at Covington & Burling.

Nathaniel O. Keohane is Director of Economic Policy and Analysis at the Environmental Defense Fund and Adjunct Professor at NYU School of Law. He has published articles on environmental economics in numerous academic journals and is the coauthor of *Markets and the Environment.*

Alexandra Khrebtukova is currently clerking at the United States Court of International Trade for the Honorable Donald C. Pogue. She represented the Republic of Palau in UN negotiations on fisheries and oceans and law of the sea, and was an IILJ scholar at NYU School of Law.

Benedict Kingsbury is Murry and Ida Becker Professor of Law and Director of the Institute for International Law and Justice at NYU School of Law. He has written extensively on trade-environment disputes, the United Nations, and interstate arbitration and the proliferation of international tribunals.

Israel Klabin is President of the Brazilian Foundation for Sustainable Development and Partner of Klabin Irmãos & Co. He worked in the Brazilian government, was elected mayor of Rio de Janeiro, and was an adviser for the Alliance for Progress under President Kennedy. He helped organize UNCED in 1992.

Rubén Kraiem is Partner and Co-chair of the Carbon Markets, Climate Change and Clean Technology practice at Covington & Burling LLP and adviser to the Coalition for Rainforest Nations.

Gabrielle Marceau is Counsellor in the Office of the WTO Director-General Pascal Lamy and Associate Professor at the Law Faculty of the University of Geneva. She has written widely on international trade and dispute settlement in international law.

Yoram Margalioth is Law Professor at Tel Aviv University and was a Global Visiting Professor of Law at NYU School of Law. He is also the Outside Director of IDB Development Corp. Ltd. He has served as Deputy Director of Harvard's International Tax Program and worked for the World Bank.

Sandra G. Mayson is a public defender in New Orleans. She received her JD from NYU School of Law, where she was an IILJ scholar. She has served as a legal adviser to Palau's Permanent Mission to the United Nations and as Publications Editor of the international environmental organization Oceana.

Bert Metz is Fellow with the European Climate Foundation and a leading member of the ClimateWorks Foundation Project Catalyst team. He led the Netherlands delegation to the Kyoto Protocol negotiations and was elected co-chairman of the IPCC's Working Group on Climate Change Mitigation for the Third and Fourth Assessment Reports.

Partha Mukhopadhyay is Senior Research Fellow with the Centre for Policy Research in New Delhi. He formerly worked with the Infrastructure Development Finance Company and was a member of the Prime Minister's Taskforce on Infrastructure. His has also worked for EX-IM Bank of India and the Trade Policy Group of the World Bank.

Sam Nader is the Head of Carbon Finance Unit at Masdar, the Abu Dhabi future energy company.

Michael Oppenheimer is the Albert G. Milbank Professor of Geosciences and International Affairs in the Woodrow Wilson School and the Department of Geosciences at Princeton University. He was a lead author on the IPCC's Third and Fourth Assessment Reports and has served on the US National Academy of Science's panels on the Atmospheric Effects of Radiation, Climate Variability and Change, and Alternative Liquid Transportation Fuels.

Annie Petsonk is International Counsel for Environmental Defense Fund. She has previously worked at UNEP and for the administrations of presidents George H. W. Bush and Bill Clinton in the Office of the US Trade Representative.

Nick Robins is Head of the Climate Change Centre of Excellence at HSBC in London. He is the co-chair of the UNEP Finance Initiative's Climate Change Working Group and a member of the UK government's Sustainable Development Panel. He has coauthored a number of reports on climate finance.

Bryce Rudyk is Research Fellow at the Center for Environmental and Land Use Law at NYU School of Law. His research focuses on financing climate change mitigation and adaptation. He has previously worked as a lawyer in transnational litigation.

Richard B. Stewart is University Professor and John Edward Sexton Professor of Law at New York University School of Law, where he directs the school's Center on Environmental and Land Use Law and Global Law School Program. He has formerly served as Assistant Attorney General for Environment and Natural Resources, US Department of Justice, and Chairman, Environmental Defense Fund.

Charlotte Streck is Founding Partner and Director of Climate Focus. She is also lead counsel for climate change with the Center for International Sustainable Development Law at McGill University and an adviser to the Prince of Wales Rainforest Project. She previously worked as Senior Counsel with the World Bank.

Murray Ward is President of Global Climate Change Consultancy. From 1996 to 2002 he led the New Zealand Ministry for the Environment's climate change team. He is a key architect of the Kyoto framework and is recognized for his international work on LULUCF and market trading mechanisms.

Jacob Werksman is Director of the Institutions and Governance Program at World Resources Institute and Adjunct Professor of law at New York University and Georgetown University. He represented the Alliance of Small Island States in the Kyoto Protocol negotiations.

Ngaire Woods is Professor of International Political Economy at Oxford University, where she directs the Global Economic Governance Programme. She has been an adviser to the UNDP's Human Development Report, a member of the Helsinki Process on global governance, of the resource group of the UN secretary-general's High-Level Commission into Threats, Challenges and Change, and of the Commonwealth Secretariat Expert Group on Democracy and Development.

Jie Yu is Director of Policy and Research at the Climate Group China. She previously worked as vice president of policy for low-carbon banking group Climate Change Capital. She is an author of China's first wind energy report and has worked on various projects to promote sustainable energy policy in China.

Part I

Climate Change and Mitigation

Overview and Key Themes

Chapter 1

Climate Finance for Limiting Emissions and Promoting Green Development

Mechanisms, Regulation, and Governance

Richard B. Stewart

University Professor and John Edward Sexton Professor of Law, NYU School of Law

Benedict Kingsbury

Murry and Ida Becker Professor, NYU School of Law

Bryce Rudyk

Research Fellow, Center on Environmental and Land Use Law

Climate finance is a critical element of global climate policy that has received far less attention than emissions limitations and climate regulatory architectures. This book redresses this deficit. It focuses on what is required to meet the need for vastly increased funding for climate mitigation and green development in developing countries. It presents new proposals to generate climate financing from both private and public sources and to deliver funds through means that will engage developing countries, build mutual trust, and secure effective long-term emissions reductions. The book also examines the vital but often neglected regulatory, trade, tax, and governance elements of global climate finance. Its proposals and analysis are designed to enrich the political and policy debate, not only for the United Nations Framework Convention on Climate Change (UNFCCC) process but more broadly. The complex issues of global climate finance cannot be resolved in a single agreement or a single forum;

they will continue to demand fresh insights and creative approaches like those presented in this volume.

1. *Three Key Determinants of Climate Finance*

Climate finance policies for limiting greenhouse gas (GHG) emissions and promoting green growth in developing countries are driven by three key sets of factors: climate science; the economics of mitigation and development needs and opportunities; and domestic and international political economy.

Climate Science Imperatives

Climate science, as set forth in the 2008 Intergovernmental Panel on Climate Change (IPCC) reports and confirmed by subsequent findings, demonstrates that we face serious risks of far-reaching climate damage unless greenhouse gas emissions growth is immediately sharply reduced. The reductions must steadily continue with the objective of stabilizing atmospheric GHG concentrations in the 450 ppmv CO_2-equivalent (CO_2e) range and thereby limiting warming to around 2°C over pre-industrial levels. (Oppenheimer, chap. 2.)

Financing Needs and Mitigation Opportunities

Even if developed country emissions are sharply curtailed, these climate targets cannot be met without very large reductions in developing country GHG emissions relative to business-as-usual (BAU) levels. Focusing on the period to 2020, a major study by Project Catalyst found that *additional* investments in developing country mitigation (over and above expected future increases in funding under existing official development assistance (ODA) programs and the Clean Development Mechanism (CDM)) in the order of €55–80 billion *each year* during the period 2010–2020 are required. A United Nations study using a different methodology estimated that the annual requirement by 2030 will be USD 92–96 billion. Significant additional amounts (estimated by Project Catalyst at €10–20 billion annually) will be needed for investment in developing country adaptation—a central issue for many African and Asian countries and small island states. We do not address it systematically in this volume because extensive further studies and innovation are required for adequate adaptation-focused financial mechanisms to be put in place. Given the limits to bilateral and multilateral ODA, which is sourced mainly in developed countries, very large amounts of private capital must be mobilized to meet the shortfall. Project Catalyst estimates that between €10–20 billion annually of private capital might be available. If this amount were used to finance mitigation actions through international credit offset markets at the market price in a single global market for all credits (with one tonne in credits for one tonne of reduction in emitted carbon-equivalents) in covered economic sectors worldwide, the reductions achieved would fall far short of that required to meet the climate targets. The conclusion is that carbon markets must be structured by governmental actions to leverage the private capital available in order to achieve significantly greater emissions reductions than would be produced by an open market, such as the current market for Certified Emissions Reduction (CER) credits issued by the CDM.

Also critical is the character of mitigation opportunities in developing countries. Project Catalyst classifies these opportunities in three broad categories based on the costs of emissions reduction. (Metz, chap. 3; Bettelheim, chap. 9.) These are

- sectors where reductions can be achieved at negative cost (i.e., mitigation investments will earn a positive economic return), mainly in energy efficiency including buildings and transportation;
- sectors where reductions can be achieved at low to moderate cost, primarily in forestry and agriculture; and
- sectors with relatively high cost reduction opportunities, primarily in energy production.

In addition, there is a need to promote low-carbon development, including through investment in infrastructure and imaginative urban policy. (Mukhopadhyay, chap. 26.)

The Political Economy of Climate Policy

As the costs of achieving even relatively modest GHG reductions, and allied concerns about international competitiveness, become politically more salient in developed countries, and as developing countries begin to confront strong demands for emissions limitations commitments, domestic political and policy factors increasingly dominate global climate policies. If the economic and political stakes continue to rise in this way, as seems highly likely, it will not be possible to sustain the UNFCCC/Kyoto model of a single universal global climate regulatory and finance regime, although it may remain a long-term goal and regulative ideal. Domestic economic and political factors in powerful states and in the European Union (EU) are increasingly setting limits to (while also motivating) inter-state agreements on climate issues. The most basic elements of global climate finance architecture must be reasonably aligned with what is politically workable within the US and the EU, accommodating also any vital points for their prosperous allies such as Australia, Canada, and Japan. Similarly, domestic policy preferences in major emerging economies such as China, India, and Brazil are part of the foundation for their positions in international climate negotiations, where they can in effect exercise a veto on many issues. The less powerful countries, both developed and developing, also have bargaining power, because unwillingness by them to vigorously follow domestic policies that are needed for various international climate agreements actually to work may blunt the purpose of the agreements and unsettle the adherence to them of the more powerful states. From the standpoint of inter-state pre-agreement bargaining and post-agreement implementation, there is what might be called a "political cost curve" in national (or regional) politics that deviates substantially from the economic cost curves that dominate in climate policy analysis. Some economically and environmentally attractive global options will not be pursued because the domestic political costs (or internal bargaining problems in the EU) would be too great, while some measures that are neither economically efficient nor environmentally optimal may prevail because they are preferred for domestic political reasons, and therefore adopted in order to achieve agreement. In principle, a global cap-and-trade system covering all countries with significant emissions, with allowance allocations to ensure equity for developing countries, would be the best solution for all if fully workable, but establishment of such an arrangement is not likely in the near term.

For political and economic reasons, both developed and developing countries are demanding greater flexibility in their international climate commitments and arrangements and greater scope to manage climate mitigation on their own terms. They are demanding latitude to take into account their different national circumstances, views of international commitments, domestic political factors, legal and institutional backgrounds, and economic costs and competitive exposures. As a result, the global climate regime has begun to move from a top-down command approach, exemplified in the Kyoto Protocol, to a more flexible bottom-up approach and assume a more plural, decentralized, and even fragmented character. (Bodansky, chap. 4.) This tendency, which while controversial has received some endorsement in the Bali roadmap and the Copenhagen process, is likely further to intensify in the coming years.

The politics of ODA in developed countries and the demands of developing countries for much greater roles in its governance will make it extraordinarily difficult to achieve a unified multilateral climate ODA mechanism with funding at adequate levels. Arrangements for global private-sector climate finance will be strongly shaped by legislation in the EU, the US, and other countries defining their markets for offset credits from developing countries. But the major developing countries, which have many lower-cost mitigation opportunities, also enjoy substantial market power. The ultimate terms of trade will likely be set through partly decentralized negotiated arrangements with many accommodations of special situations, not unlike what has occurred since 1947 under the General Agreement on Tariffs and Trade (GATT) and related trade regimes. Recipient developing countries will demand stronger commitments of both public and private funding from developed countries as the price of their participation in mitigation, and greater voice in the governance of funding mechanisms and in how funds are used. They want latitude to devise, register, and receive credits for their nationally appropriate mitigation actions (NAMAs). The challenge for climate finance will be to accommodate these various and often conflicting demands, which will generate a plurality of financing mechanisms and market arrangements, while delivering sufficient mitigation funding through means that achieve effective climate protection and green development.

2. *New Market-Based Carbon Finance Mechanisms*

The coming years will see the emergence of a variety of new climate finance mechanisms using international emissions trading markets to attract private investment in mitigation activities in developing countries. Apart from a reformed CDM, these mechanisms will generally be established pursuant to cap-and-trade regulatory systems in developed countries that recognize international credit offsets. Ideally, they should be designed to support and not retard the future adoption by major developing countries of emissions caps.

Emissions Trading Systems, Not GHG Taxes

There has been considerable debate over whether GHG emissions taxes (including carbon taxes) or a cap-and-trade system, supplemented by offset credit trading, should be used as the basic regulatory tool for limiting GHG emissions. Powerful policy and political considerations show that trading systems are superior to taxes. Caps focus political attention on environmental objectives and have the potential to ensure that they will be met. The option of issuing allowances gratis rather than auctioning them may be critical in gaining political support for climate regulation without sacrificing efficiency or effectiveness. In the international context, developing countries would never agree without compensation to impose the same level of taxes as developed countries. This would result either in differences in tax levels, creating serious leakage and loss of competitiveness in developed countries, or in the need for compensatory financing by massive transfers of ODA from developed countries. Use of international trading with generous allowance allocations to enlist developing countries is politically more feasible and more efficient in achieving mitigation.[1] Trading systems have already begun to dominate. The EU is operating a cap-and-trade system with international offset credits, the US is poised to adopt such a system, and many other developed countries will likely follow suit. (Keohane, chap. 5; Batchelder, chap. 34.)

A Plurality of Market-Based Climate Finance Mechanisms

The plural character of the emerging global climate regime will require diverse new climate finance mechanisms to accommodate the differing circumstances and objectives of both developed and developing countries. Because of the dominance of emissions trading systems for climate regulation, the inclusion of international credit offsets in developed countries' domestic legislation, as well as the CDM and its successor(s), the mechanisms for private investment will generally involve some form of climate/carbon markets. These markets will not, however, arise spontaneously, nor will they operate autonomously; they must be created, structured, regulated, and governed in order to meet the objectives of developed countries, developing countries, and investors and to protect the climate. The suite of potential climate finance mechanisms using private investment includes the following:

A REFORMED AND EXPANDED CDM

Even harsh critics of the CDM—who complain of maladministration; lack of environmental integrity in credits; failure to tap energy efficiency, renewable energy, and forestry and land use mitigation opportunities; and failure to promote long-term sustainable development—accept that some successor version of the CDM will still be needed to provide private climate finance for the least developed countries. Others believe that the CDM can be reformed so that it continues to play an important, if no longer predominant, climate financing role. The proposed reforms include changes in its governance, strengthened administrative capacities, mechanisms to promote accountability to non-state actors, steps to enhance the environmental integrity of CDM credits, removal of barriers to programmatic CDM projects, and removal of limitations on forestry, agricultural, and land-use projects. (Streck, chap. 6).

SECTORAL APPROACHES

Major developing countries have refused to assume economy-wide caps, of the type envisaged in the Kyoto Protocol model, in part because of the risk of crimping their economic development. This refusal, coupled with the limitations of the project-based CDM, has sparked wide interest in sectoral agreements under which internationally tradable offset credits would be awarded for limitations achieved in a given economic sector such as electric power generation or cement manufacture. One promising version of this approach is sectoral no-lose targets (SNLTs), under which the host developing country receives credits if it succeeds in reducing sector emissions below the target (typically set by negotiation and expressed either in terms of absolute emissions or emissions intensity) but assumes no obligations and suffers no consequences if it fails to do so. Other sector-based modalities include technology-based emissions limitations, NAMA crediting, and cooperative ventures between developed and developing country industries including technology sharing. (Ward, chap. 7.) Sector-specific targets reduce risks of unnecessarily limiting growth and better address competitiveness issues, although they of course fail to deal with emissions in sectors not covered by agreements.

Sectoral crediting, however, poses the important and investment-deterring problems that arise when one (or more) of several individual mitigation actions within the sector fails, with the result that the overall sectoral target is not fully met. From a private investor standpoint, two solutions are proposed. Host governments could indemnify participants with successful projects for any credit shortfalls. Alternatively, they could devise sector programs that specify each participant's share of the reductions needed to meet targets; credits would be awarded to those participants who achieve their share of reductions even if others do not. (Kraiem, chap. 8.)

CREDIT TRADING SYSTEMS FOR FORESTRY AND AGRICULTURE

Project Catalyst analysis reveals abundant relatively low cost mitigation opportunities in forestry and agriculture. Nearly half of the developing country mitigation opportunities during the period to 2020 fall into these categories, but most of them are not eligible for CDM credits due to CDM restrictions on these sectors. Belated recognition of these opportunities has generated proposals for forestry credits. Reducing emissions from deforestation and forest degradation (REDD), a prominent example, would award internationally tradable credits to countries that reduce historical deforestation rates. The US Waxman-Markey climate legislation envisages large volumes of credits for forest sector mitigation in developing countries. However, more is needed to sustain existing forests than just reducing deforestation rates, and the agriculture sector continues to be neglected. In order to succeed, forestry and agriculture crediting programs must recognize that a large portion of emissions are driven by the struggle of the rural poor to survive. Programs must alter the economics of rural land use, and must ensure that economic benefits from trading actually reach the rural poor. The failures of extractive industries to respect and confer sufficient benefits on local people, resulting in violence and bitter poverty in resource-rich areas, provide warnings and lessons for foreign climate mitigation initiatives based on basic changes in developing country resource uses. Such projects and policies must also promote investment in sustainable methods of intensified agricultural production as the planet's land area per person shrinks and demand for food increases. Implementing forest and agriculture offset credit systems will also require ODA and capacity building assistance to strengthen host country administrative and legal capabilities. (Bettelheim, chap. 9; Klabin, chap. 10.)

Steps to Leverage Private Investment Funds and Enhance Climate Benefits

In order to meet climate targets, market-based climate finance mechanisms must achieve robust net global emissions limitations; the Kyoto Protocol–CDM fails to do so because reductions achieved in developing countries are offset by higher emissions by developed country sources using offset credits to avoid making otherwise required reductions. The climate finance regime must also leverage the capital available; the CDM does not because it issues credits one-to-one for reductions. The requirements for net reductions and leveraging might be met in a number of different ways, although the proposals all face difficulties. (Metz, chap. 11; Petsonk, chap. 12.)

- Credits can be discounted by awarding less than one tonne of credit for each tonne of reductions.
- Developing countries may be required (for example, in sectoral crediting agreements) to achieve reductions on their own before beginning to earn credits.
- Different trading markets can be established for different types of mitigation activities, grouped by their costs per unit of emissions reduction. One market could be established for low cost energy efficiency investment, a second for higher cost forestry and agriculture investment, and a third in still higher cost energy production investments. By reducing the rents that lower cost mitigation investments would otherwise earn in a single trading market, market segmentation can stretch available capital to achieve greater reductions. A related approach is to award different levels of credits per unit of emissions reduced, with more credits in sectors in which emissions reduction costs tend to be higher.
- An international intermediary institution (or institutions) such as a "Carbon Bank" would buy, through a reverse auction or negotiated agreements, offsets from developed country suppliers at prices based on their costs and sell them to developed country credit buyers at global credit market prices. The bank would use its purchasing power to eliminate or reduce the rents that suppliers would otherwise earn by selling credits through an open global market, and thereby obtain additional reductions that could be devoted to reducing net global emissions.

- Environmental Defense Fund's CLEAR (Carbon Limits + Early Actions = Rewards) proposes adoption by developing countries of a multi-year absolute emissions limit covering either the whole economy or the major emitting sectors, establishing a Clean Investment Budget (CIB). (Petsonk, chap. 12.) This limit would initially be set at a level above its current emissions levels in order to accommodate economic growth, but below BAU. The country would earn internationally tradable allowances based on the extent to which its future emissions are below the CIB limit. Through arrangements with international financial institutions and otherwise, the allowances could be leveraged, for example by using them as collateral for debt financing for NAMAs to promote higher levels of mitigation and green development.

These mechanisms would, by one means or another, achieve leverage by reducing the amount of economic rents that developing countries would otherwise earn under open market systems. For that very reason, they will be strongly opposed by developing countries, but developed countries are increasingly likely to insist on leveraging as a condition of access to their trading markets. If the volume of credited mitigation investments increases substantially as a result of domestic legislation in developed countries, developing countries may still regard this as a gain relative to the status quo.

Linking Climate Finance Markets

The development, through a more or less decentralized process, of different climate finance mechanisms, different domestic cap-and-trade systems, and associated international allowance and offset markets will generate a variety of credit trading markets governed by different rules. In order to enhance market efficiencies and thereby achieve greater climate benefits, the different markets should be linked to facilitate cross-market trading—this will in turn require that incompatible design features be minimized. (Derwent, chap. 13.) The most important of these features are the relative stringency of caps (i.e., price paths); offset credit recognition rules (both qualitative and quantitative restrictions); the degree of long-term regulatory certainty (including the extent of potential market intervention by government); price controls (floors or ceilings); banking and borrowing rules; and the monitoring, reporting, and verification (MRV) and enforcement regime. Allowance allocation, coverage, point of regulation, and a host of other system features have no or minimal effect on the ability to link different markets. Finally, successful linking cannot occur until a pedigree of maturity and demonstrated effectiveness has been achieved in both. Private trading entities—including brokers, investors, financial services firms, and exchanges—can achieve a measure of harmonization through standard contract terms and private standard-setting mechanisms, but some of the most important features will be fixed by governments in domestic legislation. Multilateral agreements and institutions may define some key parameters, but top-down standardization of many of these features through multilateral agreements is unlikely to be feasible for some time, so harmonization of these aspects will depend in significant part on regulatory coordination among governments, partly facilitated by international institutions.

Regulation and Governance of Climate Finance Markets

Climate finance markets are neither spontaneous nor autonomous. While privately constituted or self-regulated markets are possible with regard to some specific aspects, in practice many aspects of regulation needed for climate finance markets require state action. Key features of such markets must be established and structured pursuant to domestic legislation and agreements among countries. They must be regulated to ensure that the interests of the various participating and affected countries are met, and also that climate protection and green development objectives are achieved, including through capital leveraging. At the same time, regulatory certainty on mid- to long-term targets and the implementing framework is necessary in order to attract investment capital on favorable terms. (Brinkman, chap. 14; Robins and Fulton, chap. 15.) These competing demands present vitally important but neglected issues of governance. The CDM governance issues that have only belatedly received wide recognition will be posed many times over, albeit in different institutional contexts, as new market-based climate finance mechanisms are established. These governance issues require much greater attention when new mechanisms are established, rather than postponing the problems until many years later, as happened with the CDM. The governance arrangements for these institutions include Global Administrative Law procedures for transparency, participation, reason-giving, and review in order to promote accountability and responsiveness to the various constituencies, including investors and environmental and social NGOs, with an interest in their decisions.[2]

Beyond Markets

Markets alone will not spur realization of all or anywhere near all of the relevant available developing country mitigation opportunities. In some cases, prescriptive regulation or direct government investment will be required. Moreover, even where market-based incentives can operate in ways that facilitate environmental protection and green development, they often need to be complemented and supported by other measures. For example, Project Catalyst analysis points to positive economic returns on investments in energy efficiency, but the fact that many of these theoretically profitable investments are nonetheless not being made indicates the presence of powerful institutional, informational, principal-agent, and other barriers that markets by themselves cannot overcome. Overcoming these barriers in order to enable markets to function will require host governments to take regulatory, informational, capacity-building, and other measures that will in turn depend on ODA and other support from developed country governments and multinational bodies. In other cases, the returns provided by market-based climate finance mechanisms will not be sufficient to support needed mitigation investments. These situations may require government guarantees, up-front financial support, or market support measures such as feed-in tariffs for renewable energy. (Brinkman, chap. 14; Robins and Fulton, chap. 15.) A final example is the need for long-term investment plans and policy structures to achieve low-carbon development in areas such as transportation infrastructure and urban development. (Mukhopadhyay, chap. 26.) Markets may not be capable of delivering and coordinating the required investments on the scales required. Host governments, backed by ODA and international financial institutions, will have to take a lead role, with private capital (including that leveraged from international trading mechanisms) playing a supporting role. The need for these various non-market elements underlines that developing and developed country governments and international financial institutions must play a major role in the design and governance of a climate finance mechanism using private capital.

3. Bringing Developing and Developed Countries Together in an Effective and Equitable Climate Finance System

While there is much variation, overall there is a deep lack of trust between developing and developed countries on climate change issues, and particularly on climate finance. This is due in part to a sorry history with regard to the negotiation and implementation of global commitments on development, climate, and institutional reform. Developing countries also see basic illegitimacy in demands that they sharply limit their GHG emissions without compensation for the role of already-rich countries in producing the historical stock of emissions that is causing warming today and for the future. Distrust by developing countries is intensified by the paucity of financial transfers made under the UNFCCC system, and by their dissatisfaction with the governance of several of the key climate finance institutions and arrangements. The legacy of distrust has helped make unlikely, at least for now, the possibility of a grand bargain on an encompassing global cap-and-trade system with equitable allowance allocations for developing countries. Instead, trust will have to be built step-by-step through cooperation on various means to fund initiatives in developing countries that simultaneously achieve mitigation and development goals, consistent with local circumstances and priorities.

With 1.4 billion people living in extreme poverty, poverty reduction must be a priority, all the more so as desperately poor people either are hardly emissions producers at all or have little choice about their actions (e.g., in burning forest wood for cooking and heat). In many cases they are vulnerable to serious adverse consequences both from climate change and from efforts to combat climate change by pressing emissions limitations on developing countries. Such limitations threaten the ability of developing countries to increase their energy supply in order to bring electricity to 1.6 billion people living without it, and more generally to bring modern energy sources to 2.5 billion people lacking access to them. (Ghosh and Woods, chap. 16.)

International Public Funding: Needs and Mechanisms

In order to engage and assist developing countries in limiting their GHG emissions without compromising economic development and poverty reduction, very large flows of funds to developing countries are required. Generating these flows while ensuring that they can and do reduce greenhouse gas emissions and promote socially and environmentally desirable development under arrangements of trust and confidence is the core of the global climate finance problem. Existing flows are grossly inadequate to the task. While there is much uncertainty, the scale of what may be demanded is suggested by the above-noted estimates of Project Catalyst that €55–80 billion annually of extra funding beyond that expected to be provided through expansion of existing programs is needed during the period 2010–2020, and of the UNFCCC that USD 92–96 billion extra will be needed annually by 2030.

Adaptation—the priority for many developing countries—is also vastly underfunded. Project Catalyst estimates that €10–20 billion per year will be required for adaptation, and the UNFCCC puts this estimate at USD 28–67 billion by 2030. Both estimates dwarf the current transfers for adaptation of perhaps USD 1 billion per year, including transfers under the UNFCCC. The CDM sets aside only 2% of investments to assist with adaptation costs through the Adaptation Fund. Significant further adaptation funding is envisaged in the Waxman-Markey US Emissions Trading System (ETS) bill, which makes 5% of the revenues received by the US government from auctioning permits potentially available for adaptation and technology transfer in developing countries. This apart, current proposals offer little prospect of attracting the massive funding and investment needed for adaptation, as this is difficult to integrate into the current or incipient global carbon finance systems. (Ghosh and Woods, chap. 16; Gomez-Echeverri, chap. 17.)

Some of the needed additional funds will necessarily be transfers from governments of wealthy countries to developing countries (ODA). Bilateral climate-oriented ODA has a strong programmatic and public-political dimension in initiatives such as Japan's USD 10 billion Cool Earth Partnership, Norway's Climate and Forest Initiative, Germany's International Climate Initiative, the European Union's Global Climate Change Alliance, and Australia's International Forest Carbon Initiative. Set-asides from ETS permit auction revenues, including the US ETS under the Waxman-Markey scheme and an expanded EU ETS post-2012, may generate much increased funding. However, past experience in this and other fields of bilateral ODA raise questions of whether the projected rates of disbursement will in fact be achieved, and whether such funds provide stable and sustained backing for ongoing projects and policies in developing countries over the longer term.

Potentially more important than direct bilateral ODA is the provision of funding through multilateral institutions, much of which is multilaterally routed ODA. The only financial resources under the authority of the UNFCCC Conference of the Parties (COP) are those managed by the Global Environmental Facility (GEF), the sole operating entity for the financial mechanism established by the Convention. Major issues arise as to maintaining the present mechanism, the role of the GEF going forward, and whether all compliance-linked funding should in the future be under the auspices of a single operating entity system. It has been strongly argued that an Executive Board should act as the new operating entity under the authority of the UNFCCC COP, and that a reformed financial mechanism should incorporate the principle of subsidiarity, so that decisions about where to apply the funding—for example, to underwrite NAMAs—are left (within broad parameters) to each country. (Gomez-Echeverri, chap. 17.) Under this vision, the governance structure would include national entities and implementation hubs that are linked to the UNFCCC system, the MRV system, and the system of compliance. (Gomez-Echeverri, chap. 17.)

The GEF allocates some USD 250 million per year for climate-related energy and transportation projects. Some multilateral funds outside the UNFCCC system are larger, particularly the World Bank's Climate Investment Funds, which exceed USD 6 billion divided between the Clean Technology Fund and the Strategic Climate Fund. The World Bank's Carbon Investment Unit is also active, purchasing credits on behalf of other entities. The modest scale of the World Bank's Forest Carbon Partnership Facility, at some USD 165 million, and the UN REDD funds of USD 35 million, reflect the slowness of the integration of forest issues into carbon finance structures, although the Waxman-Markey scheme and modifications envisaged to the CDM and the EU ETS may accelerate this. In total, these multilateral funds, even taking into account projected bilateral ODA, are nowhere near large enough for what is needed. Their objectives and policies were often formulated with very limited developing country participation. Moreover, each fund typically has separate procedural rules and its own governance structure. Many have insufficient transparency and accountability. Because of the operational complexity of many of the funds, dedicated experts are required at the national level in order to access and benefit from them, sapping the already weak national monitoring and reporting capacities of many developing countries, and imposing high transaction costs. In many cases they fund projects rather than programs or sector plans of action, limiting their ability to respond to developing country priorities in overall development strategy.

Governance of International Public Funding

Housing these funds within the World Bank or conceivably the International Monetary Fund (IMF) is the general preference of developed countries seeking assurances about strong management and prevention of misappropriation. Developing countries, however, lack effective votes and voice in these institutions (even with reform of the IMF), and resent the dominance of the industrialized countries and the effective veto power of the US. The GEF attracts similar objections, leading many developing countries to prefer it to be simply an operational entity, not a financial mechanism. The Adaptation Fund has more appeal for developing countries as a model for climate finance governance, with a Board comprising 16 members and 16 alternates representing the five United Nations regional groups (2 from each), the small island developing states (1), the least developed countries (1), Annex I Parties (2), and non–Annex I Parties (2). (Ghosh and Woods, chap 16.)

The credibility of the climate public finance regimes will be enhanced if the principal inter-governmental financing mechanisms are actually able to monitor and evaluate the effectiveness of financial flows, combining self-reporting by member states with institutional reporting of the origin and destination of financial flows. A review capacity—to assess the timeliness, adequacy, and impact of financial transfers—would buttress the system. Developing countries are also pushing for binding multilateral financial commitments from developed countries as an essential part of any global deals that would include some form of limitations commitments by major developing countries. They have proposed international agreement on means of raising additional public funds for mitigation investment in developing countries, including dedication of revenues from auctioning allowances in developed countries' domestic trading systems, taxes on international emissions trading, and international levies on bunker and aviation fuels. A much less ambitious approach would be to include funding initiatives by developed countries in the framework proposed by Korea for registering national climate undertakings, including NAMAs by developing countries.

Financing Bottom-Up Approaches to Climate Mitigation in Developing Countries

Whereas developing countries tend to favor strong participatory interstate governance of financial mechanisms, with regard to emissions controls for developing countries they generally favor bottom-up approaches, such as NAMAs, over top-down approaches, such as explicitly binding targets or systems with implicit future targets. In addition to political and equity arguments (made also by some developed countries) for greater autonomy, more specific environmental and developmental arguments are advanced for flexibility and bottom-up approaches to promote mitigation actions adapted to the circumstances (including institutional circumstances) and priorities of individual developing countries. It is argued, first, that strengthening domestic institutions in developing countries remains essential to successful low-carbon development. (Dubash, chap. 18.) Where national institutions are dysfunctional or severely distorted by capture, top-down measures such as emissions trading systems with caps or targets—designed to change relative prices, signal economic opportunity, and stimulate actors to capture efficiency—are in practice blunted and even produce distorting effects. Second, trying to generate targets for developing countries currently risks perverse results. Classifying any sectoral reforms by reference to standard cost-curve metrics and methodologies, such as negative cost, co-benefits actions, and positive cost, involves drawn-out negotiations and may be counterproductive. Such classifications give countries incentives to demonstrate that their possible actions carry high positive costs, which means they need to avoid undertaking these actions unless they receive climate financing. Thus, sectoral approaches can risk discouraging early action while rewarding stonewalling and late action. (Dubash, chap. 18.) Moreover, any approach to calculation of credits that requires construction of a counterfactual baseline (such as a business as usual (BAU) baseline) against which to judge progress, risks gaming and high transaction costs. Thus, in the short run, when early action is at a premium, a bottom-up approach to climate mitigation may well deliver more and earlier mitigation than top-down approaches. (Dubash, chap. 18.)

The bottom-up approach depends on there being both the incentives and the capability for developing countries to take significant national measures on their initiative. The Korean proposal for registration and crediting of NAMAs seeks to provide the incentives. The very concept of

NAMAs, and then the formal possibility of registering them, provides a form of international and local recognition that has helped catalyze some national action. Much greater impetus comes, however, from the possibility that NAMAs that produce emissions limitations as confirmed by MRV might receive financial support from the global climate finance regime. Financing for NAMAs may be unilateral (provided by the developing country itself, typically where there are also economic or other non-climate reasons to take the action), provided by grants or investment by foreign states or multilateral institutions (supported NAMAs), or through recognition with carbon offset credits (credited NAMAs). (Chung, chap. 19.) This proposal does not, however, solve the capability problems: the need for developing countries to have the capability to identify and implement promising NAMAs; define their emissions baselines and trends and the projected effect of a new policy or measure; facilitate the necessary measurement, reporting, and verification of the reductions; and manage any financial inflows in a responsible and accountable fashion. Some, such as Mexico, have actively built up capability and generated GHG inventories and baselines to support a substantial catalogue of prospective NAMAs. Brazil has also taken substantial steps, particularly with regard to forests and its Amazon Fund, but also in some industrial and energy sectors. But many developing countries do not have this ability or the financial, institutional, and personnel resources to build it very quickly. Capacity also depends on technology transfer in many instances. In all of these respects, effective bottom-up approaches to climate mitigation have much in common with long-standing problems in development and development assistance. Because capacity building is not itself a NAMA under any ordinary definition, ancillary arrangements for capacity building and technology transfer are essential.

Conditionality in Climate Funding

Aid donors and concessional funders of low-carbon green development or of mitigation measures unsurprisingly want to set conditions on the use of their funds, and to ensure close supervision. This raises major problems about fairness of conditions and of their construction and supervision, particularly what might be called the good governance of conditionality.

Applying some conditions to developing country performance is inevitable, and may indeed be helpful in overcoming opportunistic tendencies of some leaders and officials to divert funds for private or political ends. However, many unilateral conditions are viewed antagonistically by developing countries. In the GEF, conditionalities are set and enforced in what is perceived as a one-sided fashion through the "contributor prerogative." It is argued instead that developed countries should work in partnership with developing countries to use their investments to build institutional and policy conditions in recipient countries for more sustainable climate-related polices to take root. (Werksman, chap. 20.) Such a reciprocal deal could encompass direct access to funding with relaxed conditions for developing countries whose national institutions can demonstrate that they meet fiduciary standards through sound national systems for measuring, reporting, and verifying (MRV) funded actions. Such quality assurance and accountability mechanisms would be an integral part of a new deal on international funding for the bottom-up approach. (Werksman, chap. 20.) Indonesia's proposal that incoming funds go into its Climate Change Trust Fund for onward distribution may prove a test case for such arrangements.

Conditions are also set by private funders, such as the group of commercial financial institutions adhering to the Equator Principles, which itself integrates closely with the inter-governmental but private-sector-oriented International Finance Corporation (IFC), so that Equator banks are expected in their project lending to insist on IFC Performance Standards, even where the IFC is not a funder for the project. These and other conditions set by private financing sources increasingly incorporate climate-related requirements. But the reasons for doing so are complex, and it cannot be presumed that these conditions are cost-effective, reflect the best interests or priorities of developing countries, or are necessarily adhered to. This phenomenon of private or hybrid public-private conditionality plays an ever more visible part in climate finance, but its effects and actual significance have not yet been sufficiently evaluated. (Davis and Dadush, chap. 21.)

The politics and psychology of donating money, particularly public money, often generate strong donor-set incentives and conditions in the belief that they will lead the recipient to adopt and achieve the donors' objectives. In practice, however, such structured incentives or conditionality may often reflect other donor predilections, and they may well impede realization of the stated objectives. (Woods, chap. 22.) On the recipients' side, local ownership (including local willingness to provide resources for the project), local management and implementation, and local control of redesign and adaptation of the project as these become needed make a huge difference to success. On the funders' side, rich countries that are potentially willing to accept tough binding emissions commitments are much less willing to accept binding financial commitments. This raises uncertainties that may increase the risk for developing countries in making long-term commitments, having had much experience in the past with projects undertaken with careful adherence to a bevy of conditions, and which the donor then decides not to continue funding. (Woods, chap. 22.) Assuring financing from private markets raises other difficult complications of stability.

4. National Policies and the Global Climate Finance Regime

As well as being politically inescapable, there are many other reasons to build an international climate regime in ways that accommodate some existing and future national policy choices. Pluralism can have global policy benefits in encouraging experimentation, learning, and improvement. Allowing different national approaches may enable agreement on more demanding levels of climate mitigation and assistance. More scope is left for national political processes, including democratic processes where these function well, in making future choices. Significant deference to developing countries is demanded by them, as an acknowledgment of their sovereignty coupled with acknowledgment of their limited role in historical carbon build-up from anthropogenic emissions. These concerns can lead many developing countries strongly to resist simply accepting what appear to be instructions on climate policy from developed countries, even if the proposed policies may be entirely well-intended and accompanied by full and adequate financial support. Yet, the multiplicity of national policy approaches that the bottom-up ethos celebrates faces the hazard of being a cacophony that neither produces much climate change mitigation or forest and environmental protection nor generates cost-effective and socially beneficial development for people who need it. Some significant overarching regulation, supervision, and coordination are therefore essential. In this light, part 4 of the book focuses on some key national (and EU) policies and the interactions both among these different national measures and with an emerging international climate finance regime.

Developed Country Climate Legislation and Global Carbon Markets

As discussed above, flows of (usually private) funds made possible because investors receive carbon offset credits—which have value due to their tradability in the carbon markets of the developed countries—have considerable importance for mitigation in developing countries. Both the European ETS and the Waxman-Markey legislative scheme in the US limit the percentage of emissions permits derived from foreign offsets, and both seek to promote some offsets in their own territories. They also limit the kinds of foreign projects that can generate offset credits usable in their markets: thus, the EU excluded forest projects from the ETS, the Waxman-Markey scheme envisages excluding many projects not meeting specific US standards, and the New Zealand scheme excludes credits relating to nuclear power projects.

The Waxman-Markey scheme in the US is designed to be open to some potential integration with, but also to strongly influence, other national and international emissions abatement and carbon finance schemes. Up to USD 1 billion per year in credits from approved foreign and international cap-and-trade systems will be accepted in the US, although after a phase-in period this will be at a 20% discount. However, the foreign or international schemes will be required to meet stringent substantive and procedural standards, to be applied by US government agencies (principally the Environmental Protection Agency), an arrangement likely to require application of Global Administrative Law principles and procedures to ensure adequate consideration of the interests of other countries, other investors, and other global constituencies. This legislation also seeks to move toward sectoral crediting for certain countries and sectors over time, and will render individual projects ineligible for crediting where it would be covered sectorally. (Keohane, chap. 23.)

The EU ETS has been the main source of demand for CDM credits. Steps by the EU to toughen up on recognition of these credits is likely to force some reform of the CDM, which may raise some problems of unilateralism even as reforms are much needed. At the same time, efforts to bolster the carbon price and stability in the EU ETS market, through laying out a predictable total cap beyond 2020 and other measures such as making it an EU-wide market with auctions rather than continuing with highly variable national measures, will give support to the CDM and other offset credit systems. The EU is also taking steps to foster an eventual global ETS market, based on the expected national cap-and-trade schemes in the US, New Zealand, Australia, and elsewhere. (Chapman, chap. 24.)

Developing Countries' Initiatives and Policy Innovations

China does not (and likely for a long time will not) accept an economy-wide emissions cap. However, it is taking an increasingly significant raft of voluntary measures (often driven by economic modernization and energy security goals) which may substantially reduce emissions below BAU, while also advancing some development objectives including rural electrification using some renewable sources. The government has required increased energy efficiency in building designs and pursued reductions in emissions intensity especially in the power sector. This and other policies have driven up the demand for ultra-supercritical power stations, wind power equipment, and other technologies that due to large-scale production have dropped in price, helping to establish their Chinese manufacturers as leaders in these global markets. The possibility of registering these actions as NAMAs, and conceivably receiving credits far beyond those generated by the current range of CDM projects in China, may bring China further into the climate finance regime. (Yu, chap. 25.)

Within the complex mix of national, inter-governmental, and global policymaking structures, good climate policy innovation must be actively fostered and receive quick recognition and financing. Much of this innovation must occur in sub-national political units, such as cities. While US cities typically use much more energy per capita than European or other cities, the variance among US cities is very large, and comparable variance is beginning to appear amongst Chinese cities. Some of this can be redressed through building standards and other transposable initiatives, but much relates to complex combinations of historical development and current policies concerning the role of public transport, tax and other incentives to live densely or diffusely and close or far from work, as well as some cultural conditioning. (Mukhopadhyay, chap. 26.) Reform of urban policy might have major emissions-reducing effects: perhaps one-third of emissions mitigation in India by 2050 could be through lower-carbon cities. But it is not readily incentivized or funded through private investments driven by crediting for the major foreign offset markets. Urban policy is so complex that it must be tailored to innumerable local specificities and political structures—making metrics, replication, and rapid diffusion difficult—and it must necessarily be pursued largely though bottom-up processes.

All of this calls for further reflection on what drives national policy formation on climate issues. The US and EU political processes have received intense study, so the factors influencing the approaches emerging there are broadly understood even if not robustly predictable in their outcomes; but much less is generally known about Chinese policymaking processes. An interesting experiment potentially related to future policy formation is the Masdar initiative to create a moderate-sized carbon-neutral city with innovative technology in Abu Dhabi, which if it succeeds could conceivably be an incubus for rethinking national and international approaches to climate change in several oil-exporting states with high per capita emissions and incomes. (Nader, chap. 27.)

Understanding the Evolution of National and Global Climate Policies

In none of these cases is the national government (or the EU) forming policy in an entirely autochthonous fashion, even if the national processes can seem quite insular. Each takes some account of policies elsewhere, of positions in international institutions, and of some broad global forces and trends. In this respect, a model of a two-level game, in which national officials and interest groups act in national politics and in inter-governmental politics, is insufficient. Some elements of both national and inter-state policy formation on climate issues extend beyond simply interest-driven bargaining. In some part, the politics is global, at least in the modest sense of being not simply national or inter-governmental, as the work of the IPCC or of major transnational climate lobby groups illustrates. National policies are also shaped by processes of mimesis or diffusion. A few basic models of cap-and-trade credit offset carbon market design and regulation may emerge, as existing national schemes are studied by the next adopters. Best practices may also develop, on matters ranging from treatment by national electricity regulators of renewable supplies to the grid (e.g., through feed-in tariffs) to certification and verification of emissions reductions. Such standardization may potentially facilitate both financial flows and regulatory design.

Autonomy in national or regional climate policies may indeed be an objective of some who wish to maintain the possibility of national control (or patronage and rent-seeking), but it comes at a high cost in unrealized efficiency gains. A proliferation of regulatory arrangements invites arbitrage and opportunism that may eventually lead to the ironing out of incongruities, but at considerable fiscal and environmental cost. Regulatory competition likewise can have benefits, but also major costs. Regulatory cooperation, mutual recognition arrangements, and real coordination between national regulators and funders with different objectives and constituencies may become effective only very slowly. Some structures of transnational and international regulation will almost inevitably be demanded, but will come into tension with the values of bottom-up approaches. Such tension is already manifest in questions concerning the application of global trade law to climate issues, and may develop in the future on some taxation issues affecting climate finance.

5. *Trade Law and Climate Policies*

Climate finance and regulation and international trade law will increasingly intersect. As international and, more pertinently, national climate change regulations affect and potentially distort trade between states—not only between states that adopt GHG emissions regulation and those that do not, but also between states that adopt differing levels and forms of regulation—international trade law will be implicated. (Marceau, chap. 28.) Potential or actual World Trade Organization (WTO) challenges to domestic climate measures (and similar challenges under regional trade agreements) might chill or retard the implementation of domestic climate regulation. But trade law may also have a positive influence on the design of measures to combat competitive and leakage concerns, as well as prevent protectionism in the guise of environmental measures. Climate measures will also test the limits and analytical precision of the environment-related exceptions under Art. XX of the General Agreement on Tariffs and Trade (GATT) and similar exceptions in other WTO agreements. Because the issues likely to arise are complex and novel, the impact of the multitude of trade rules on climate finance and mitigation are difficult to anticipate and address. However, WTO officials, at least, are optimistic that the WTO agreements can accommodate properly designed domestic climate regulatory measures.

Trading Climate Assets

While the trading of Assigned Amount Units (AAUs) between Annex I states is regulated by the UNFCCC and Kyoto Protocol, trading across borders and systems of allowances issued under domestic cap-and-trade systems and other assets created pursuant to climate regulatory law, such as renewable energy certificates (RECs), is not explicitly addressed in WTO agreements or any other current international agreement. It is likely that the WTO would have some jurisdiction over this trading and government measures to regulate or support the market, but it is not clear whether allowances will be treated as financial instruments or other types of services under the General Agreement on Trade in Services (GATS), or potentially as goods under GATT. Similar uncertainties arise in relation to offset credits produced through the CDM and joint implementation under the Kyoto Protocol and under the trading systems created pursuant to domestic cap-and-trade systems in the EU, US, and other developed countries. Because of the nature of the transactions involved, which might be seen as investments with government involvement, the provisions of the Agreement on Trade-Related Investment Measures (TRIMS), the Government Procurement Agreement, or the Technical Barriers to Trade (TBT) Agreement might apply as well as GATS and GATT. (Marceau, chap. 28; Howse and Eliason, chap. 29.)

Border Measures to Address Leakage and Competitiveness Issues

There is strong political concern that climate regulation will impair the competitiveness of firms and sectors in regulated economies relative to those in states with less stringent or no regulation. Because investment and business activity will tend to flow to jurisdictions with lower production costs, difference in domestic climate regulations will, absent countervailing international or domestic rules, result in leakage of production emissions to jurisdictions with weaker or no regulation. The result is not simply a loss in economic competitiveness in regulating jurisdictions (which threatens domestic political support for climate regulation), but a loss of environmental effectiveness, as the emitting activities are shifted around rather than reduced. Moreover, leakage spurs carbon-intensive development in jurisdictions with weak or no regulation, making it more difficult for them to reverse course in the future. International agreement on common climate regulatory policies is one solution. In its absence, states may well adopt domestic rules requiring imported products be accompanied by emissions certificates like those required of domestic producers under domestic cap-and-trade laws, or be subject to some form of economically equivalent border carbon credit adjustment. (Khrebtukova, chap. 31.) The effect is to impose an economic charge reflecting climate externalities on all goods, whether domestic or imported, consumed in the regulating jurisdiction. States, including developing countries, which regard climate externalities as less costly and oppose strong regulations, will of course oppose carbon levies on their exports. Although the issues of trade regulatory law are again complex and novel, border carbon measures may well be consistent with WTO rules if applied in an evenhanded way without discrimination against imported goods. Adoption of such measures by some states will spur their adoption by others, creating a bottom-up pattern of international regulation that may eventually provide a foundation for international agreement on common climate regulatory norms.

Free Allocation of Climate Assets and Direct and Regulatory Climate Subsidies

Another step that regulating states may take to protect their industries' competitiveness is to issue emissions allowances for free rather than auctioning them. In most of the current and proposed developed country cap-and-trade systems, all or most of the allowances are distributed gratis at least for the short- and mid-term. (Keohane, chap. 23; Chapman, chap. 24.) The WTO Subsidies and Countervailing Measures (SCM) Agreement contains specific rules concerning subsidies and limits to them where they may cause adverse effects on trade. Under one interpretation of free allowance allocations to domestic producers—as a transfer of a valuable asset from the government to private entities without compensation—they and tax breaks with similar effects might represent actionable or countervailable subsidies under WTO law. An analogous logic might conceivably conclude that states that do not regulate their carbon emissions when a majority of states do so are granting their industries an unlawful subsidy under the SCM. (Howse and Eliason, chap. 30.) Direct subsidies—whether for production or export—for climate-friendly technologies, including tax breaks and other forms of direct government financial support for wind, solar, and biofuels, as well as regulatory measures such as feed-in tariffs and renewable energy portfolio and credit standards, also pose issues under the SCM Agreement; in the case of biofuels, the Agreement on Agriculture is also relevant.

Carbon Footprint and Other Standards Created by Non-state and Hybrid State-Private Actors

The proliferation of initiatives for carbon footprint labeling schemes currently being developed by business and non-profit organizations alone and also in conjunction with states could adversely affect developing country exports and pose international regulatory and governance concerns. Mandatory carbon labeling standards adopted by states, as Japan contemplates, would be subject to potential challenge for failure to conform to the TBT Agreement's Code of Good Practice for standard setting. It remains an open question whether these requirements apply to privately run labeling schemes that have some form of state sponsorship or involvement.. (Mayson, chap. 33.) Alternatively, states may adopt as mandatory private carbon labeling standards and invoke them as "relevant international standards" which, under the TBT, create a "safe harbor" presumption of legality when the state rules are challenged. It is unclear whether and under what circumstances private voluntary standards might enjoy such a presumption, including where there are competing private standards. The legal validity of carbon footprint labeling standards can be strengthened if the initiatives are based on widely accepted scientific and standard-setting principles, adopted with adequate transparency and broad-based participation, and accompanied by technical assistance to developing countries and small producers to support compliance.

Developing Country Concerns with Climate-Related Trade Measures

Developing countries are concerned by developed country motivations in climate policy generally, and especially so as regards the move to link trade measures with climate. (Ghosh, chap. 32.) One concern is that climate-related trade measures such as border carbon adjustments will be used for protectionism and eco-imperialism camouflaged as environmental protection. Developing countries are also concerned that current steps to lower barriers against trade in environmental goods and services (under negotiation in the Doha round) could be implemented in a lopsided way that disadvantages developing countries. A further concern is that stringent intellectual property rights may inhibit needed technology transfer. To prevent unjustified trade distortions and potential inequities, it is argued that better reporting by states of relevant domestic trade measures is needed, along with greater capacity in the WTO and in developing countries to monitor domestic trade measures, and greater transparency in climate-related domestic initiatives that impact trade. (Ghosh, chap. 32.)

6. Taxation Issues in Climate Finance

The tax treatment of emissions trading systems (which as discussed above are the dominant instrument for achieving mitigation) and the new types of assets (emissions allowances and offset credits, collectively "permits") that they create is an important subject just beginning to achieve recognition. Tax issues are important because the efficiency and effectiveness of trading systems in achieving climate protection goals can be seriously compromised by inappropriate domestic tax policies and by international differences in tax treatment.

Emissions trading markets produce cost savings and enhance environmental benefits relative to traditional prescriptive regulation because they allocate emissions limitations among sources in the most cost-effective pattern, and thereby achieve aggregate limitations at lowest cost. Trading systems achieve this efficient result because sources seeking to minimize their overall costs of dealing with emissions will invest in emissions abatement to the point where marginal abatement costs equal the cost of acquiring or continuing to hold permits, which is the same as the market price of permits. Since, in a given trading system, the market price of permits is the same for all sources, their marginal abatement costs will also be the same, producing an efficient abatement allocation. (This explains why it is desirable to link different trading systems so that sources covered by different systems all face the same permit price.)

The tax treatment of abatement costs and of permits can impair regulatory efficiency by disrupting the equilibration of marginal abatement costs and permit costs. For example, a country may grant tax subsidies to certain politically favored emission abatement technologies, such as ethanol or wind power, thereby reducing their after-tax costs. As a result, more investment will flow to such technologies and less to other abatement methods that, pre-tax, have lowers costs, undermining the efficiency of the trading system and driving up the overall costs to society of limiting emissions. Similar distortions and inefficiencies can occur in the international allocation of abatement investments if different countries adopt different tax rates for abatement or for permits. The resulting inefficiencies may not only create very large amounts of economic waste, but also undermine political support for strong climate mitigation regulation by driving up abatement costs. Analysis of these tax issues leads to the following conclusions (Batchelder, chap. 34; Kane, chap. 35; Margalioth, chap. 36):

If an emissions trading system is adopted, tax and other subsidies for particular abatement methods or for energy use should be, to the maximum extent feasible, eliminated unless justified by non-climate externalities, because they threaten to create market distortions, regulatory inefficiencies, and economic waste.

Distortions and regulatory inefficiencies caused by differences in the tax treatment of abatement and permit costs can arise either within a given jurisdiction or across jurisdictions. The major source of problems will be the persistence (contrary to the immediately above policy recommendation) of tax and other subsidies for particular abatement methods, such as renewable energy. Two different strategies can be used to eliminate or reduce the resulting distortions. First, tax all permit costs the same across all jurisdictions, and also tax all abatement costs the same across jurisdictions; if this is achieved, it is not necessary also to equalize the treatment of abatement and permit costs within any jurisdiction. Second, tax all permits and abatement costs the same (at the margin) in each jurisdiction; if this is achieved, it is not necessary also to equalize tax rates and other tax rules among jurisdictions. As a practical matter, it is much less difficult to implement the second strategy than the first. This strategy is compatible with tax and other subsidy programs for certain specific abatement methods if they are properly designed. International agreement by major states on adopting this strategy should be pursued through multilateral climate negotiations rather than bilateral tax treaties.

Distortions and inefficiencies can also be independently caused by the various aspects of the tax treatment of permits that create a lock-in effect that leads firms to hold permits longer than they otherwise would in order to defer taxes on the increased value of the permits. As a result, permit values will rise because of tax considerations, distorting the tradeoff between abatement and holding permits. Partial solutions include auctioning permits or taxing the value of gratis permits when issued. Tax changes should also be adopted to address distortions caused by the interaction between fluctuating permit prices and tax rules.

Differences in the treatment of abatement costs and of permit costs in different jurisdictions will require tax authorities to develop transfer pricing rules to police tax arbitrage practices by multinational businesses operating in more than one jurisdiction that pose risks of trading market distortions.

Finally, trading systems present important macro-level issues of efficiency and equity. By imposing a cost on emissions, trading systems increase the price of energy and of goods and services produced by it, which has a net regressive effect. Auctioning permits and using the proceeds to make direct transfers to lower-income households or providing them with tax credits can offset or reduce this effect.

7. Conclusion: The Ways Forward on Climate Finance

The issues raised by climate science, economic analysis, and the political economy of climate policy, fleshed out in the chapters of this book, generate rich and powerful implications for future carbon finance arrangements. These include the following:

- A variety of new arrangements to generate public and private climate finance and engage developing countries in mitigation are needed; a single uniform design is neither feasible nor desirable. Ideally, they should be designed to support and not retard the future adoption by major developing countries of emissions caps.
- A suite of revised or new market-based mechanisms must be developed to mobilize very large increases in private investment in developing country mitigation. These include a reformed CDM and credit offset trading systems established pursuant to domestic cap-and-trade climate regulation by developed countries.
- These mechanisms must leverage private investment in order to achieve net climate benefits and secure long-term low-carbon development.

- Carbon markets cannot be autonomous; they must be structured, regulated with developing as well as developed country involvement in their design and governance. Governance arrangements should be transparent and provide for appropriate mechanisms for accountability to non-state actors including investors and NGOs.
- Linkages among national and regional regulatory/trading systems through allowance trading and transfers of offset credits will play a key role; achieving them will require coordination among governments.
- Governance arrangements and the determination of conditions on ODA must be changed significantly to enhance developing countries' roles, build trust, and assure climate-sustainable development. Greater integration or coordination of international ODA mechanisms is also needed.
- The new arrangements for both private investment and ODA must be structured to match with the different types and costs of mitigation opportunities available in developing countries.
- New institutional arrangements are needed to recognize, facilitate, and coordinate the diversity of decentralized climate initiatives among both developing and developed countries.
- WTO trade rules need to be interpreted and applied to accommodate domestic climate-related regulatory measures, including border carbon adjustments to deal with competitiveness and leakage issues and mitigation technology subsidies, so long as they are nondiscriminatory and not protectionist.
- The WTO and developing countries need to develop additional capacities to monitor and respond to adoption of climate-related domestic measures that impact trade in potentially distortionary or protectionist ways.
- Changes in tax laws, including a degree of harmonization among national tax systems, are needed in order to avoid creating market distortions and regulatory inefficiencies in trading-based climate regulatory systems.

NOTES

1. Richard B. Stewart and Jonathan B. Wiener, *Reconstructing Climate Policy: Beyond Kyoto* (2003).

2. The Global Administrative Law Project at New York University School of Law undertakes and promotes academic research and policy debate on the use of these mechanisms to improve global regulatory governance. See www.iilj.org/gal. An overview of global administrative law is provided in Benedict Kingsbury, Nico Krisch and Richard B. Stewart, "The Emergence of Global Administrative Law," 68:3–4 *Law and Contemporary Problems* (2005), 15.

Chapter 2

Understanding the Causes and Implications of Climate Change

Michael Oppenheimer

Albert G. Milbank Professor of Geosciences and International Affairs, Woodrow Wilson School and the Department of Geosciences, Princeton University

Key Points

- Carbon Dioxide (CO_2)—emitted through electricity generation, transport, agriculture, and forestry—is responsible for four-fifths of the warming effect of current emissions of long-lived greenhouse gases and will persist in the atmosphere for many decades, with a significant fraction remaining for more than a millennium. CO_2 levels are already higher than any time in at least the past 850,000 years.
- While the effects of climate change cannot be predicted with certainty because future emissions trajectories are not known and our understanding of the climate system (particularly feedbacks) is limited, we are already seeing significant climatic impacts, including: increasing mean ocean temperature and sea level; increasing extremes of heat and drought; changes in ranges of species; melting of ice sheets, Arctic sea ice, and glaciers; and increasing severity of some extreme climatic events.

Causes of Climate Change

The basic scientific framework of the climate change issue is well understood: greenhouse gases (GHG) emitted in the process of electricity generation, transport, agriculture, and forestry are accumulating in the atmosphere, gradually altering the heat balance of the Earth and inevitably changing its climate. The greatest concern arises from long-lived gases (carbon dioxide, methane, halocarbons, and nitrous oxide) because they persist in the atmosphere for a period ranging from decades to longer than a millennium after release. Of these, carbon dioxide is the most important because it accounts for about four-fifths of the warming effect of current emissions of the long-lived GHGs. Atmospheric carbon dioxide levels are already one-third greater than in preindustrial times, and higher than at any time in at least the past 850,000 years. Other trace constituents emitted from human activity affect the climate in important ways, but are much less persistent. These include ozone (a key component of smog) and soot and other particles, the latter having both warming and cooling effects.

All this we know with certainty. It is also certain that over the past century, the Earth has warmed by about three-fourths of a degree Celsius (°C). It is very likely that the combined influence of all these gases and particles has caused most of the observed warming of the past half-century.

Carbon dioxide from the combustion of fossil fuels (coal, oil, and natural gas) for electricity generation, transport, and other purposes produces almost 60% of the warming effect of the current emissions of long-lived gases. Another 20% comes from carbon dioxide and other gases emitted during the cutting and burning of forests for the purposes of conversion of lands for timber production, agriculture, pastoral use, and related settlement. Climate change cannot be slowed significantly, and the climate cannot be stabilized, without large reductions in emissions from fossil fuels and strong measures to curb deforestation.

Consequences of Climate Change

There are two general sources of uncertainty in projecting future climate change. First, estimates of future emissions of the greenhouse gases vary widely, although most projections envision emissions continuing to grow for at least the first half of this century. The second source of uncertainty arises from our limited understanding of the climate system, particularly the responses (called feedbacks) of the individual components of the

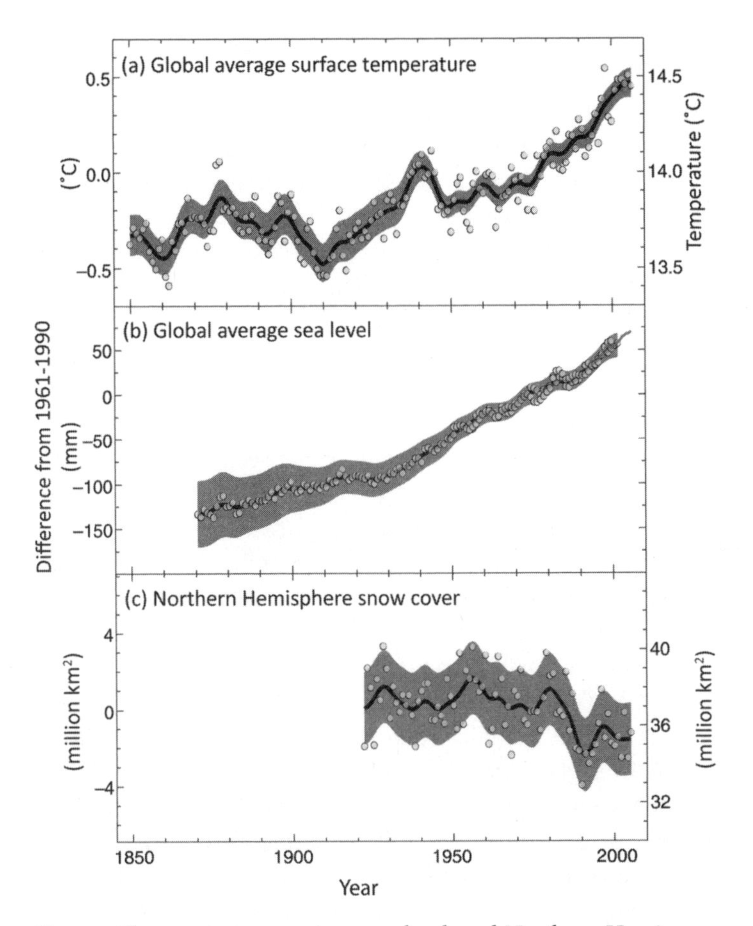

Fig. 2.1. Changes in temperature, sea level, and Northern Hemisphere snow cover. Observed changes in (a) global average surface temperature, (b) global average sea level from tide gauge and satellite data, and (c) Northern Hemisphere snow cover for March–April. All differences are relative to corresponding averages for the period 1961–1990. Smoothed curves represent decadal averaged values, while circles show yearly values. The shaded areas are the uncertainty intervals estimated from a comprehensive analysis of known uncertainties (a and b) and from the time series (c). (Source: *Climate Change 2007: Synthesis Report; Contribution of Working Groups I, II, and III to the Fourth Assessment Report of the Intergovernmental Panel on Climate Change,* Figure SPM.1, IPCC, Geneva, Switzerland)

Earth system—including clouds, ice sheets, and ocean circulation—to the initial greenhouse warming. The range of possibilities is enormous.

If prompt action is taken to stem emissions, it remains possible that a modest additional global warming of not much more than 1°C would occur. Even if limited to this level, such warming would be greater and faster than any global climate change during the history of civilization, and would doubtless cause disruption of ecosystems and risk of extinction of some species, as well as problems for many nations, especially developing countries in coastal or semi-arid regions. On the other hand, unconstrained emissions would lead to a warming that could reach as high as six degrees, which would present us with an unmitigated worldwide disaster.

Either of these scenarios, and any in between, would be expected to result in intensification of all of the current climate trends. Atmospheric warming has already resulted in a mean ocean temperature increase of nearly 0.8°C. Polar ice sheets in Greenland and Antarctica are shrinking at their peripheries. Summer Arctic sea ice is retreating, opening navigation routes around the North Pole. The 2007 Report of the Intergovernmental Panel on Climate Change (IPCC) estimates that a global warming of about 3–4°C by 2100 (in the middle of the projected range) would cause the Arctic to become largely free of summer ice, while more recent estimates suggest this outcome could occur before midcentury. The oceans are becoming more acidic as they absorb some of the carbon dioxide added to the atmosphere. The resulting effects are likely to translate into increased difficulty for shell-forming organisms, like coral, and substantial effects on marine ecosystems, food chains, and all those that depend on them, including humans.

With a somewhat lesser degree of certainty, we can say that extremes of heat and drought have increased. When precipitation does occur, there is a tendency for it to fall with greater intensity, increasing the potential for flooding. The IPCC indicates that a 3–4°C warming and associated drought probably would significantly reduce agricultural productivity in developing countries in the tropical and subtropical regions, where malnutrition and episodic starvation are already endemic. Of particular concern is the potential reduction of water available on the Asian subcontinent as Himalayan glaciers shrink, with the outcome that some of the major rivers, including the Ganga, may maintain significant flow only seasonally. Extreme heat waves of the sort that struck Western Europe in 2003 —associated with the deaths of at least 35,000 people—would become the

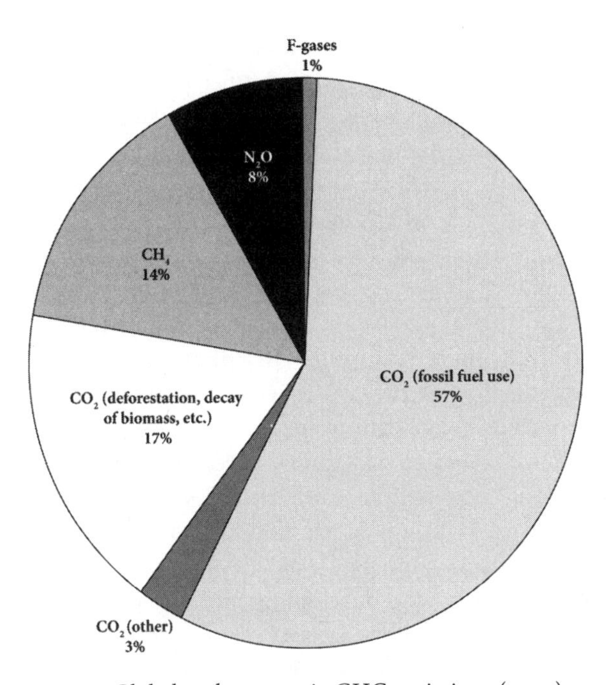

Fig. 2.2. Global anthropogenic GHG emissions (2004). (Source: *Climate Change 2007: Mitigation of Climate Change; Working Group III Contribution to the Fourth Assessment Report of the Intergovernmental Panel on Climate Change,* Figure TS.1b, Cambridge University Press)

norm rather than a rare occurrence, and even more extreme events are expected to occur. While human ability to adapt to such impacts may improve over time, it is likely that many other species will fail to adjust successfully. The IPCC estimates that 30% or more of all species will become at risk of eventual extinction at a persistent warming below 3–4°C.

Perhaps the broadest threat from a geographic perspective relates to the projected rise in sea level. IPCC's projection, a rise of 18–59 cm over this century, accounts for two of the three major drivers of sea level rise: expansion of ocean water and melting of glaciers. However, it does not fully account for the potential contribution from ice sheets because, at the time, there was no satisfactory way to do so. But over the past two years, a variety of preliminary estimates of how large the contribution from the ice sheets of Greenland and Antarctica may become have appeared in the literature, resulting in a possible total sea level rise of as much as 1–2 meters during this century, with a further multi-meter increase during the remainder of the millennium. Such a sea level rise would devastate wetlands; obliterate many low-lying, densely populated deltaic areas, including much of Bangladesh; and wreak havoc along coastlines in the developed world as well, where monumental amounts of permanent infrastructure would be at risk, forcing a costly (if gradual) retreat. A sea level rise of this sort appears to have occurred in the distant past when Earth warmed to similar levels, but at that time fixed human settlement had not yet evolved and retreat would have been far easier.

A close examination of the full range of potential impacts indicates that the most serious risks begin to increase markedly once warming exceeds 1–2°C above recent temperatures. Based on such findings, the EU has adopted a long-term objective of limiting warming to no more than 2°C above recent temperatures (corresponding to about 1.2°C above preindustrial temperatures). This goal was endorsed by the major emitting countries, both developed and developing, meeting in July 2009 at an unusual joint conference held at the annual G-8 meeting.

The opportunity to avert such a warming shrinks markedly with every year of further delay in reducing emissions. Of particular concern is the rapid growth in emissions from large developing countries like China, India, and Brazil. Unless developed countries are able to reduce their emissions substantially over the coming decade as a first step, and unless developing countries are able to lower their emissions significantly below business as usual expectations during the following decade, there is little chance that such a warming would be averted.

Responses to Climate Change

With these concerns in mind, we should quickly develop and implement policies and institutions (both internationally and domestically) to rapidly change our carbon emissions trajectory and provide the means to cope with the inevitability of some additional warming. These include:

1. Institutions and financing that would facilitate adaptation—already a key issue—even in developed countries.
2. Policies that would effectively impose a continuously increasing price on carbon. Such policies must include a stringent cap in the

2020 timeframe, along with subsequent reductions on emissions from all developed countries. The US, Canada, Australia, Japan, and many European countries have yet to act to reduce their emissions.
3. A collaborative decision on the part of countries with large emissions on the respective roles and responsibilities of developed and developing countries in achieving emissions limitations, along with adoption and implementation of a treaty that embodies these concepts in specific numerical obligations, accompanied by enforcement provisions and appropriate financing mechanisms. Rapid agreement on reduction of deforestation is an important supplement to limitations on fossil fuel emissions.
4. Funding and collaborative arrangements sufficient to provide incentives for research and development, and commercialization of emerging low-carbon technologies.

These objectives offer a stark challenge requiring immediate and focused attention by governments. An honest reading of the scientific evidence provides no excuse for hesitation. Prompt and effective action to reduce emissions is our only option.

FURTHER READING

Intergovernmental Panel on Climate Change, Fourth Assessment Report (2007), Full Report available at http://www.ipcc.ch/pdf/assessment-report/ar4/syr/ar4_syr.pdf, Summary for Policymakers available at http://www.ipcc.ch/pdf/assessment-report/ar4/syr/ar4_syr_spm.pdf.

Chapter 3

The Climate Financing Problem

Funds Needed for Global Climate Change Mitigation Vastly Exceed Funds Currently Available

Bert Metz

Senior Fellow, European Climate Foundation

Key Points

- Even assuming ambitious GHG reductions by developed countries, large additional reductions in developing country emissions will be required in order to limit global warming to 2°C. This pathway requires global emissions to peak no later than 2015, and to fall 50% from 1990 levels by 2050, split so that developed nations shoulder the majority of the burden.
- In developing countries, some of these reductions have negative costs, such as energy efficiency in buildings, transport, and industry. Many areas have moderate positive costs (agriculture and forestry), and technology-intensive sectors (notably renewable energy) require significant funding.
- On the basis of the principle of compensation for incremental costs by developed countries, a total of €65–100 billion annually over the 2010–2020 period is needed to finance these reductions and meet developing countries' adaptation needs. However, these cost figures do not capture the significant positive externalities throughout society from low-carbon investment such as increased employment, heightened energy security, improved agricultural productivity, and improved infrastructure.

Background

The latest assessment of the Intergovernmental Panel on Climate Change (IPCC) clearly shows that climate change risks will be manageable if global mean temperatures do not increase more than 2°C above the pre-industrial period. This requires a global trajectory towards stabilization of greenhouse gas (GHG) concentrations in the atmosphere of 450 ppmv CO_2 equivalent (CO_2e) to give us even a 40–60% chance of meeting the 2°C target. This requires global GHG emissions to start declining no later than 2015 and fall to 50% below 1990 levels by 2050. For the period ending in 2020, this translates into a global emissions reduction of 17 Gt CO_2e compared to business as usual (BAU) by 2020 (see Figure 3.1).

Existing technologies can achieve over 90% of the global emissions reductions needed by 2020. Technology costs are already rapidly declining, and new technologies will further reduce costs and increase effectiveness. The costs of low-carbon transition are manageable. If the savings from negative cost mitigation actions can be effectively captured through intelligent regulation and incentives, the costs of more expensive investments

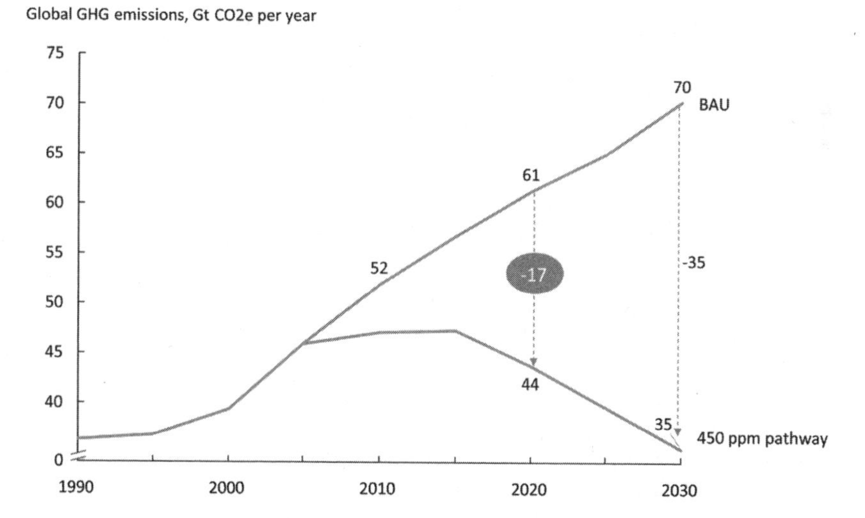

Fig. 3.1. 17 Gt of reductions below the reference pathway in 2020 are required to stay on a 450ppmv pathway. (Source: McKinsey Global GHG Abatement Cost Curve v2.0 (2009); M. G. J. Den Elzen and M. Meinshausen, *Multi-gas emission pathways for meeting the EU 2°C climate target,* 2006; IEA World Economic Outlook 2007; Project Catalyst analysis)

can be offset. The main question of this essay is, "what level of financing will make achieving these reductions possible?"

Developed and Developing Country Contributions

Equity demands that developed countries need to realize substantial emission reductions by 2020 of 25–40% below 1990 on average (with differentiation amongst them). We do not have the luxury of time to enter into a global climate agreement where developed countries move first and developing countries follow on behind. Developing countries need to deliver the rest of the reductions in order to meet the overall global emissions freeze and decline. According to scientific analysis, developing countries' emissions should be 15–30% below the BAU baseline by 2020. The question is, how this can be realized in a way that is consistent with the negotiation mandate that was agreed upon in Bali in December 2007 (the Bali Action Plan), and that is fair to developing countries with their generally low incomes and limited responsibility for current climate change?

Project Catalyst assumes that developing countries implement their contribution in the form of a low-carbon development plan—made up of nationally appropriate mitigation actions (NAMAs)—that steers their economies towards a low-emission, sustainable economy over a longer period of time through specific NAMAs. This ensures that climate change mitigation is a development-oriented transformation of the economy that would enable countries to avoid large negative impacts from further climate change. It would also have many benefits for energy security, health, employment, mobility, and competitiveness.

The Funding Needed by Developing Countries

Based on this notion of low-carbon development, estimates have been made of the incremental costs of capturing the opportunities for energy efficiency improvement in buildings, transportation, and industry; moving to a low-carbon energy supply and reducing deforestation; improving sustainable forest management; and moving to sustainable agriculture. Figure 3.2 shows the McKinsey cost curve for the group of developing countries. Costs of measures are expressed in euro per tonne of CO_2e avoided, based on social rates of return (4%). These costs are the costs for the society, not the costs for private investors.

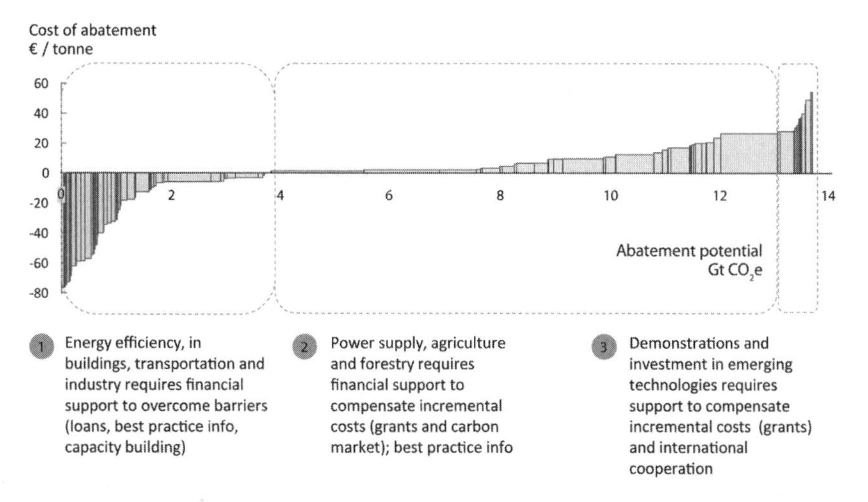

Fig. 3.2. Different financial support for different areas of the cost curve. Developing country abatement cost curve, 2020 (up to costs of €60/t). (Source: Project Catalyst analysis; McKinsey Global GHG Abatement Cost Curve v2.0 (2009))

The curve shows many opportunities (approximately one-third of the required reductions) with negative costs, meaning they pay for themselves because of saved energy costs, mostly in buildings, transportation, and industry, with an average rate of return on investment of 17%. For the agriculture and forestry sector, most options have moderate positive costs. Power sector costs are generally higher. Some emerging technologies, such as solar PV and concentrated solar power, have even higher costs, given their current state of development.

Investment in all of these sectors—especially the second—also has a strong record of stimulating growth across the economy through similar historical analogies (railroads and electrification, for example) and recent data on green job creation and its positive effects on society, and these benefits are not fully borne out by the cost curve above. These benefits include increased energy security, reduced energy prices and volatility in the long term, reduced vulnerability to energy price shocks, and reduced pollution from particulates.

Based on this cost curve, the total incremental cost (i.e., the total of all positive cost measures) for developing countries can be calculated. The negative costs are not subtracted because in most cases government policies and measures are needed to capture the negative cost potential; these will require substantial action from developing countries and even international support in the form of capacity building or loans to overcome up-front capital constraints.

Adding up the incremental costs for the period 2010–2020 gives an average total of €35 billion per year. Allowing a higher rate of return in developing countries and covering transaction costs and specific funding for emerging technologies brings the total to €55–80 billion annually. To this total, the incremental costs of adaptation measures in developing countries need to be added. Catalyst estimates these adaptation costs at €10–20 billion per year on average for the period 2010–2020, just for knowledge development, disaster management, and planning, with significantly more after this timeframe. This brings the overall amount of funding needed to support developing countries in making their contribution to an ambitious Copenhagen agreement and adapting to climate change to €65–100 billion per year (see Figure 3.3).

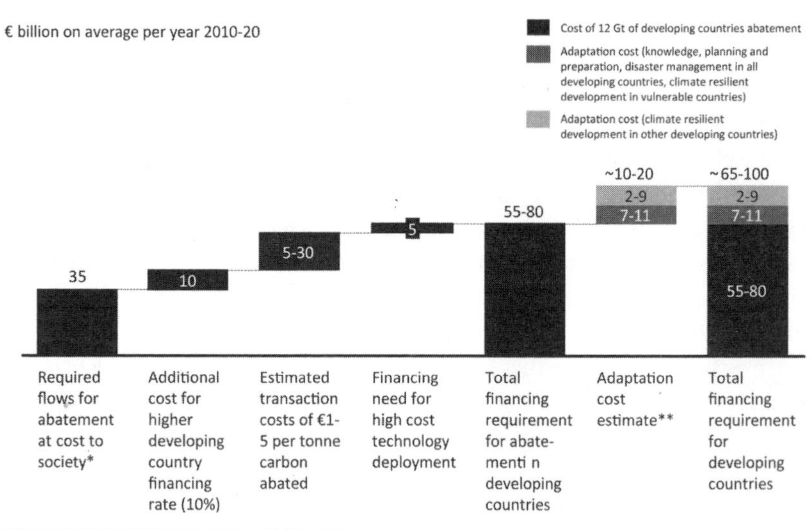

Fig. 3.3. Developing countries would require up to €65–100 billion per year in incremental cost financing. (Source: McKinsey Global GHG Abatement Cost Curve v2.0 (2009); V. Bosetti et al. "International energy R&D spillovers and the economics of greenhouse gas atmospheric stabilization," *Energy Economics* 30(6) (2008); UNFCCC; Project Catalyst analysis)

FURTHER READING

Intergovernmental Panel on Climate Change, *Climate Change 2007: Synthesis Report; Contribution of Working Groups I, II, and III to the Fourth Assessment Report of the Intergovernmental Panel on Climate Change* (2007), available at http://www.ipcc.ch/ipccreports/ar4-syr.htm.

McKinsey and Company, *Pathways to a Low Carbon Economy* (2008), available at https://solutions.mckinsey.com/climatedesk/CMS/Default.aspx.

Netherlands Environmental Assessment Agency, "Chair's Summary Report: Where Development Meets Climate: Development Related Mitigation Options for a Global Climate Change Agreement," available at http://www.pbl.nl/en/dossiers/Climatechange/Publications/International-Workshop-Where-development-meets-climate.html.

Project Catalyst, *Towards a Global Climate Agreement: Synthesis Briefing Paper June 2009*, available at http://www.project-catalyst.info.

Chapter 4

The Future of Climate Governance
Creating a More Flexible Architecture

Daniel Bodansky
Emily and Ernst Woodruff Professor of International Law,
University of Georgia School of Law

Key Points

- To ensure greater participation, it is essential to allow greater flexibility for states to mitigate climate change on their own terms.
- National mitigation actions will need to be integrated into an international agreement to ensure environmental effectiveness.
- As the recent proposals from the United States and Australia suggest, flexibility in deciding on climate commitments is not just a concern of developing countries.

Everyone wants to learn from history, so as not to repeat it. But what are the lessons of the Kyoto Protocol? Although opinions differ widely, a growing consensus accepts the need for greater flexibility in a new climate change agreement. The Kyoto Protocol targets cover only about one-quarter of global emissions. Perhaps the central challenge for a new climate agreement is to broaden this coverage by getting the United States, China, and other major emerging economies on board. Although giving states greater flexibility in their choice of commitments will not, by itself, be enough. However, it could make a new agreement more attractive to a wider group of states by allowing them, in setting commitments, to take into account their differing national circumstances, views of international commitments, domestic political processes, legal backgrounds, and economic costs.

Flexibility in the Context of Climate Change

International agreements vary widely in the latitude that they give participating countries. At one end of the spectrum, some agreements take a uniform top-down command approach, requiring states to undertake particular policies and measures. At the other extreme, agreements can adopt a highly flexible bottom-up approach, allowing each of the participating states to define its own commitments. In the environmental realm, the Convention on International Trade in Endangered Species (CITES) illustrates the top-down approach. It prescribes which species to protect and how to do so (through a permitting system for imports and exports). Similarly, the international oil pollution treaty (MARPOL) prescribes very specific rules regarding the construction and design of oil tankers. Conversely, the US-Canada Air Quality Agreement illustrates a bottom-up approach, codifying in an international agreement the pre-existing domestic air pollution programs of the two participating states.

When it was adopted, the Kyoto Protocol was hailed by many for its flexibility. Rather than requiring states to adopt particular policies and measures such as efficiency standards, the Kyoto emissions targets give states freedom in deciding how to reduce emissions and (to a limited degree) when and where to do so. But although Kyoto gives states freedom in deciding how to *implement* their commitments, it does not give them similar flexibility in *defining* their commitments. Instead, it prescribes a single type of international commitment (absolute emissions targets relative to a fixed historical baseline), the scope of those targets (economy-wide), the gases covered (a basket of six greenhouse gases), and the international offsets that can count towards meeting the targets (certified emission reductions created through the collective decisionmaking of the Clean Development Mechanism (CDM)). As a result, states that are worried about the risks to economic growth posed by an absolute, economy-wide emissions cap, or that wish to focus on a particular sector or gas, or that prefer to adopt a price-based rather than a quantity-based instrument (that is, a tax rather than a quantitative cap on emissions) are effectively excluded from the regime.

Flexibility in the choice of commitments is particularly important in the climate change regime because of the huge domestic sensitivities involved—much greater than the sensitivities raised by any prior international environmental issue. Climate change implicates virtually every area of domestic policy, including industrial, agricultural, energy, transportation, and land-use policy. Building domestic coalitions to address the problem will require many compromises (as the drafting of US climate change legislation currently illustrates). A new international climate agreement needs to encourage states to do more, but it also needs to give states the necessary space for their domestic political processes to unfold. The importance of flexibility has long been recognized for developing countries in articulating nationally appropriate mitigation actions. But, as the United States and Australian proposals in the Copenhagen negotiations emphasized, it is also of concern to developed countries.

A Growing Consensus

The need for greater flexibility was a central conclusion of the Climate Dialogue at Pocantico, a group of policymakers and stakeholders from 15 countries convened by the Pew Center on Climate Change. As the Pocantico report explained, "the types of policies that can effectively address greenhouse gas emissions in a manner consistent with national interest will by necessity vary from country to country. To achieve broad participation, a framework for multilateral climate action must therefore be flexible enough to accommodate different types of national strategies by allowing for different types of commitments. It must enable each country to choose a pathway that best aligns the global interest in climate action with its own evolving national interests."[1]

A Flexible Approach: The US and Australia Proposals

What might a more flexible approach entail? The United States' proposal for an implementing agreement suggests one option.[2] It envisions developed countries committing to emissions targets, but allows them to implement their commitments "in conformity with domestic law."[3] Although the meaning of this phrase is not altogether clear, it appears to allow developed countries, through their national legislation, to specify their targets in somewhat different ways. Of course, for the international targets to have any determinate meaning, there must be limits to these national variations. But, within reasonable bounds, a new climate regime should recognize the reality that developed countries may decide to define their targets differently in their national legislation—for example, with respect to precise sectoral coverage, base years, or allowable offsets.

A potentially broader type of flexibility is illustrated by an Australian proposal to establish schedules of national commitments and actions, which is similar to the nationally appropriate mitigation action (NAMAs) registry proposal of Korea.[4] Rather than defining commitments through a top-down negotiating process, as in Kyoto, states would engage in a bottom-up process, in which they would develop national schedules of commitments and actions and then register those commitments and actions internationally. As the Australian proposal explains, the schedule approach would "give Parties substantial flexibility to craft commitments and actions in a manner appropriate to their national circumstances." Schedules could include both legally binding commitments as well as non-binding actions. The Australian proposal suggests that developed country schedules should include comparable mitigation efforts, including emission targets, while developing country schedules could include other types of commitments or actions, such as sectoral targets or particular policies and measures.

Balancing Flexibility and Effectiveness

As both the US and Australian proposals recognize, in providing for greater flexibility, it is important to retain elements of integration in the new regime. A system of pledge and review, in which each state merely comes forward with its own national programs, would be extremely flexible, but it would not produce a sufficient level of effort. States may be unwilling to put forth their fullest effort unless they are confident that those efforts will be reciprocated by others at a roughly comparable level. Although states should have a certain degree of flexibility in their choice of commitments and actions, these commitments and actions need to be negotiated together and integrated into a single international regime, to promote reciprocity and coordination of national efforts.

To the extent that states undertake different types of commitments and actions, this will make the task of ensuring the comparability of efforts among countries even more challenging and urgent than under an exclusively targets-based approach. In the Bali Action Plan negotiations, states have proposed a wide array of criteria to assess the comparability of developed country commitments. These include: the form and nature of commitments (legal vs. non-legal, quantified vs. unquantified); their comprehensiveness and duration; a country's absolute and per capita levels of emissions, emissions reduction potentials, geography, resource endowment, economic structure, and historical responsibility; and provisions for third-party review and compliance.[5] Although agreement on a common methodology or formula to assess comparability of efforts seems unrealistic, much more analytical work is needed to enable countries to make their own individual assessments of one another's efforts in order to reach a politically acceptable outcome.

Conclusion

Is breaking the impasse on climate change merely a matter of elaborating a more flexible architecture? Obviously not. Flexibility is a necessary but not a sufficient condition for agreement. States first must have the political will to act. The point of flexibility is to avoid creating obstacles to agreement, so that, when states do decide to act, they have the freedom to move forward in a manner that makes sense for them.

FURTHER READING

Daniel Bodansky and Elliot Diringer, *Towards an Integrated Multi-Track Climate Framework* (Pew Center on Global Climate Change, 2007).

Pew Center on Global Climate Change, *Report of the Climate Dialogue at Pocantico* (2005).

NOTES

1. *Report of the Climate Dialogue at Pocantico* (Washington, DC: Pew Center on Global Climate Change, 2005), p. 9.
2. "US Submission on Copenhagen Agreed Outcome," UN Doc. FCCC/AWGLCA/2009/Misc.4 (Part II), p. 106.
3. Id., art. 2.1(a).
4. "Schedules in a Post-2012 Treaty," Submission of Australia, UN Doc. FCCC/AWGLCA/2009/Misc.4 (Part I), p. 22.
5. *See* Revised Negotiating Text, FCCC/AWGLCA/2009/INF.1, paras. 55–59.

Part II

Proposals for Climate Finance

Regulatory and Market Mechanisms and Incentives

A

Trading or Taxes?

Chapter 5

Cap-and-Trade Is Preferable to a Carbon Tax

Nathaniel O. Keohane

Director of Economic Policy and Analysis, Environmental Defense Fund

Key Points

- Contrary to the views of many economists and policy analysts, cap-and-trade systems are superior to taxes for limiting GHG emissions.
- The key advantage of cap-and-trade over a carbon tax is that a cap puts a direct limit on the quantity of emissions, while letting the market determine the price. This ensures not only that the environmental objective is met, but also that the political and social debate around a cap-and-trade program is appropriately focused on environmental goals.
- Cap-and-trade facilitates international harmonization and cooperation on climate policy, thereby reducing the costs of limiting emissions on a global basis. Cap-and-trade easily accommodates linkages between national emissions trading systems that will in turn equalize the marginal cost of abatement across these countries. Cap-and-trade can also provide incentives for developing countries to reduce emissions in order to gain access to carbon markets in developed countries.
- Proponents of carbon taxes often criticize cap-and-trade because it typically involves free allowance allocations. But from both an economic and environmental perspective, how allowances are allocated is less important than the stringency of the cap. Further, the allocation of allowances fulfills a potential political function in building support for a system that puts a price on carbon.

Background

First proposed in 1968, cap-and-trade came into its own in 1990 with the passage of the US Clean Air Act Amendments, which created an emissions trading system for sulfur dioxide emissions from electric power plants. That program has cut emissions in half at less than a third of the predicted cost, with overwhelming benefits to human health and ecosystems.[1] Since then, the European Union has established its Emission Trading Scheme (ETS) for carbon dioxide (CO_2) to achieve its emissions targets under the Kyoto Protocol. And cap-and-trade is the centerpiece of climate legislation under consideration in the US Congress. Despite these successes, calls for a carbon tax still abound. This chapter compares the two policy instruments and argues that a cap-and-trade system is superior for controlling greenhouse gas (GHG) emissions, on political, policy, and economic grounds.

Under a cap-and-trade program, total allowable emissions are limited (the cap), and an equivalent number of allowances are created, which may be bought or sold on a market (the trade). At the end of each compliance period, each regulated facility must submit allowances in an amount equal to its emissions. In many systems, firms may also bank allowances for use in later years, or borrow them in limited amounts from future periods.

Both a cap-and-trade system and a carbon tax put a price on carbon —giving polluters strong economic incentives to reduce pollution cost-effectively, and creating a powerful reward for technological innovation. Both policy instruments also take advantage of the information available to individual agents, rather than relying on the limited knowledge of regulators to identify and mandate facility-level performance requirements or the use of specified technologies.

In theory, if the marginal costs of abatement are static and known with certainty, and in the absence of any political considerations, a carbon tax and cap-and-trade program can be designed to be perfectly equivalent with respect to the allocation of abatement across firms, the marginal price of emissions, and the real economic costs of achieving a given emissions target.[2] A cap-and-trade system and a carbon tax are also similar in terms of administrative costs (the costs of operating a trading market are relatively minimal), and require the same amount and accuracy of emissions data to monitor and enforce compliance.

Raising Revenue

Despite their theoretical similarities, cap-and-trade and a carbon tax exhibit important differences in practice. One commonly cited distinction is that a tax raises government revenue, while cap-and-trade programs have typically involved generous allocations of free emission allowances to regulated entities. (Note that while free allocation is a common feature of cap-and-trade programs, it is not a necessary one: emission allowances could alternatively be sold at auction, raising the same expected revenue as a tax.)

From an economic perspective, whether the government raises revenue is less significant than it may appear. What matters more is how the economic value represented by the allowances is allocated. Economic efficiency can be enhanced by auctioning allowances (or imposing taxes) to raise revenue and using it to reduce pre-existing distortionary taxes on labor and capital[3]—but only if politicians are willing to reduce marginal tax rates rather than spending the revenue on per capita rebates or government programs.

If revenue is not recycled so as to reduce such distortions, how the economic value in allowances is allocated has implications for distributional incidence—but not for efficiency. Free allocation, by itself, does not undermine the environmental or economic performance of a cap-and-trade system: that performance depends on the incentives created by the allowance price, which is a function of the stringency of the cap rather than the method of allowance allocation. On the other hand, free allocation is a powerful political tool, offering a ready means of calibrating the trade-off among different interests. This political flexibility is likely to make cap-and-trade more effective than a carbon tax in accommodating political realities while still accomplishing the ultimate goal of controlling GHG emissions.

Price vs. Quantity

A much more fundamental difference between cap-and-trade and a carbon tax is the distinction between setting a price and controlling quantity. Under a cap-and-trade program, the total quantity of cumulative emissions—and thus the environmental performance of the program—is fixed. The price of emissions is uncertain, however: it is generated by the allowance market, determined by factors such as the stringency of the cap, the pace of technology development, the prices of fossil fuels, and energy demand.

In contrast, a carbon tax determines the price directly, but leaves actual emissions uncertain—dependent on factors such as the rate of economic growth, the cost and availability of abatement technology, and policies adopted in other jurisdictions (which set the international terms of trade). As a result, a tax may not achieve—indeed, is unlikely to achieve—any particular level of cumulative emissions specified in advance.

This prices vs. quantities distinction is important for several reasons. First, the goals of climate policy are commonly defined in terms of quantity targets: temperature changes, GHG concentrations, or cumulative emissions. While some economists have advocated setting a price equal to the marginal damages from emissions, we simply lack the necessary information to do so. A recent survey by the Intergovernmental Panel on Climate Change (IPCC) found that estimates of marginal damages vary by a factor of 30, from USD 3 to USD 95 per metric tonne of CO_2, and that many of those estimates ignore non-market damages and catastrophic impacts.[4]

Second, framing the issue in terms of price or quantity leads to very different debates about policy objectives. A proposal to tax emissions focuses the debate on the size of the tax and the potential costs to the economy. In contrast, a proposal to cap emissions frames the discussion in terms of emissions targets and the consequences of climate change. As a consequence, cap-and-trade is likely to lead to more ambitious emissions reduction goals, while a tax is tantamount to proposing a less stringent policy.

Third, cap-and-trade enhances the prospects for harmonizing international action. Averting dangerous climate change will require deep cuts in GHG emissions by the world's advanced economies as well as meaningful reductions from middle-income and developing countries. Doing so at the lowest possible cost, however, requires that the marginal costs of abatement be equated across countries. In this context, cap-and-trade has a key advantage over a carbon tax: marginal costs can be equalized simply by linking allowance markets in different countries. Individual countries or regions that establish domestic cap-and-trade programs can let regulated firms purchase allowances from other systems for compliance with their own, with minimal coordination. In contrast, achieving cost-effectiveness through a harmonized carbon tax requires explicit international agreement upon a common tax rate, an enormous political challenge and most likely unattainable.

Fourth, a cap-and-trade system can promote broad international participation. Carbon markets in the developed world will be a powerful attractor for emerging economies. These countries, which are rich in low-cost abatement opportunities, would be net sellers in a global carbon market —giving them a strong economic incentive to join. (Carbon markets thus serve the goal of equity as well as efficiency, providing a scalable means of financing low-carbon development.) Leading developing economies ready to take on domestically enforceable targets could take full advantage of carbon markets by linking their own cap-and-trade systems with those in developed countries. Other developing countries, lacking the capacity to establish cap-and-trade systems in the near term, could participate by selling offset credits. In turn, the EU and US will have considerable leverage to push for strong action on climate change, in return for carbon market access. An emissions tax provides neither such an incentive nor such leverage.

Price vs. Quantity and Economic Efficiency

Although the arguments just outlined favor cap-and-trade, the distinction between price and quantity instruments also provides what is typically cited as the strongest economic argument for a carbon tax. When marginal abatement costs are uncertain, the relative efficiency of a price instrument versus a quantity instrument depends on the relative slopes of the marginal benefit and marginal cost functions. Because the marginal benefits of reducing GHG emissions are generally thought to be flat relative to the marginal costs, many economists have concluded that a tax will be preferable on efficiency grounds.

This argument hinges on the presumption that marginal benefits of abatement are flat—equivalently, that the harm from emitting a ton of greenhouse gases stays roughly the same as emissions increase. This depends in turn on two (often hidden) assumptions: first, that the relevant policy problem is one of managing the flow of emissions (for example on an annual basis); second, that policies are path-independent—in other words, that the initial choice of policy does not constrain subsequent policies. Neither assumption is valid. Actual cap-and-trade programs for

GHG emissions would allow full banking and borrowing, setting a cumulative target rather than a series of annual targets. This approach is well-suited to climate change, where impacts are driven not by short-term emissions but by the accumulation of long-lived greenhouse gases in the atmosphere. Moreover, the difficulties and fixed costs involved in passing legislation, and the gradual adaptation of regulated entities to established policies, mean in the real world that those policies will be politically difficult to change. Once a framework is put in place, it is likely to remain. These arguments suggest that the problem ought to be defined in terms of concentrations (or cumulative emissions over several decades), and that policies should be assessed in light of their performance over a similar time horizon.

Once the problem is defined in terms of cumulative emissions under a long-lived policy framework, the nature of the damages from climate change takes on new significance. Growing scientific evidence suggests that climatic responses to temperature increases are highly nonlinear and characterized by tipping points—levels of warming that would trigger relatively rapid and irreversible changes in major components of the Earth system. Examples include the loss of Arctic summer sea ice, the melting of the Greenland and West Antarctic Ice Sheets, the weakening of the North Atlantic Thermohaline Circulation, loss of coral reefs, and the disappearance of the Amazon rainforest. These nonlinearities in the damages from climate change imply that the marginal benefits of abatement, far from being flat, may be relatively steep, when measured in terms of cumulative emissions. The intuition is simple: When the climate system exhibits threshold effects, and policies are hard to change once enacted, putting a limit on cumulative emissions is preferable to setting a price, in order to ensure that we don't exceed dangerous tipping points.

Although the precise temperatures at which these thresholds occur are admittedly uncertain, such uncertainty compounds the concerns rather than alleviating them. We cannot rule out the possibility that we are headed for truly catastrophic consequences: Weitzman, for example, estimates that there is a 5% chance that business-as-usual emissions will lead to a warming of more than 10°C and a 1% chance of exceeding 20°C. The overwhelming importance of such fat tails in the probability distribution of harms diminishes the significance of the expected (average) welfare maximization framework that underlies the prices vs. quantities argument, which fails to give adequate weight to relatively low probabilities of very serious harm. Instead, what Weitzman calls his "generalized precautionary principle" dovetails with the more general argument that climate policy is best viewed in terms of risk management. Even if these arguments do not (yet) provide a theoretical argument for cap-and-trade over a carbon tax, they lend urgency to the practical arguments made here: namely, that a cap-and-trade program is a more promising approach to achieve the near-term emissions reductions needed to hedge the risk of catastrophe.

Conclusion

This chapter has presented the case for using cap-and-trade, rather than a carbon tax, to control greenhouse gases. A system of tradable permits offers a great deal of flexibility in allocating the value of emissions, enhancing its political feasibility. Cap-and-trade also promotes cost-effectiveness, broad participation, and equity in the international context, with much less coordination than a tax would require. Finally, controlling the cumulative quantity of GHG emissions is likely to be superior to setting a tax even on narrow economic efficiency grounds, given the importance of limiting GHG concentrations below potentially dangerous thresholds in the climate system.

FURTHER READING

Nathaniel O. Keohane, "Cap and trade rehabilitated: Using tradable permits to control U.S. greenhouse gases," *Review of Environmental Economics and Policy* 3: 1–21 (2009).

Gilbert E. Metcalf, "Designing a carbon tax to reduce U.S. greenhouse gas emissions," *Review of Environmental Economics and Policy* 3 (2009).

William D. Nordhaus, *A Question of Balance* (New Haven, CT: Yale University Press, 2008).

Robert N. Stavins, *Proposal for a U.S. Cap-and-Trade System to Address Global Climate Change: A Sensible and Practical Approach to Reduce Greenhouse Gas Emissions,* Hamilton Project Discussion Paper 2007-13. (Washington, DC: Brookings Institution, 2007).

Gary Yohe, Natasha Andronova, and Michael Schlesinger, "To hedge or not against an uncertain climate future?" *Science* 306: 416–17 (2003).

NOTES

1. National Acid Precipitation Assessment Program, *National acid precipitation assessment program report to Congress: An integrated assessment* (Washington, DC: National Acid Precipitation Assessment Program, 2005).

2. W. David Montgomery, "Markets in licenses and efficient pollution control programs," *Journal of Economic Theory* 5: 395–418 (1972).

3. Lawrence H. Goulder, Ian W. H. Parry, and Dallas Burtraw, "Revenue-raising vs. other approaches to environmental protection: The critical significance of pre-existing tax distortions," *RAND Journal of Economics* 28: 708–31 (1997).

4. G. W. Yohe et al., "Perspectives on climate change impacts and sustainability," in M. L. Parry et al., eds., *Climate Change 2007: Impacts, Adaptation, and Vulnerability; Contribution of Working Group II to the Fourth Assessment Report of the Intergovernmental Panel on Climate Change* (Cambridge, UK: Cambridge University Press, 2007), pp. 811–41.

B

Reforming the Clean Development Mechanism (CDM)

Chapter 6

Expectations and Reality of the Clean Development Mechanism

A Climate Finance Instrument between Accusation and Aspirations

Charlotte Streck

Director, Climate Focus

Key Points

- The CDM has, by many accounts, met its objective in terms of the funds it has leveraged from the private sector to achieve mitigation in developing countries, the capacity it has built, and the awareness it has raised, not to mention the lessons it has provided.
- Despite these successes, the CDM has been roundly criticized from many fronts in terms of its governance practices, environmental integrity, and contribution to sustainable development.
- The CDM has too much experience and future potential to justify abandoning it in the post-2012 climate framework. Much needed reform, focusing on improving the environmental and administrative credentials of the scheme and an expansion of its scope and scale, will transform the CDM into a truly useful tool for sustainable development and climate policy.

Introduction

Born in the last hour of the Kyoto Protocol negotiations with modest expectations, the Clean Development Mechanism (CDM) offers a story of unprecedented success. By June 2009, the CDM Executive Board (EB) registered more than 1,500 projects that are expected to create 1.6 billion tons of greenhouse gas (GHG) emission reductions by 2013. The CDM has attracted the interest of the private sector in industrialized and developing countries alike and built a global carbon market.

The CDM initiated a paradigm shift in support of developing country action under multilateral environmental treaties. In its design, negotiators relied heavily on experience from the Global Environment Facility (GEF) and the Multilateral Fund for the implementation of the Montreal Protocol. They modeled the EB after the Multilateral Fund's Executive Committee, and introduced the concept of additionality, closely related to the incremental cost principle of the Multilateral Fund and the GEF. At the behest of the US, negotiators however introduced two innovations in the CDM's design, making its operational character fundamentally different from those of the GEF and the Multilateral Fund: (i) investment was linked to tradable emission certificates; and (ii) private entities authorized by State Parties were invited to participate. By involving markets and private actors, the Kyoto Protocol leveraged significant financial resources for low-carbon investment in developing countries. In 2007 and 2008 alone, the CDM mobilized USD 15 billion in primary transactions in Certified Emissions Reductions credits (CERs). In comparison, the GEF —the single biggest environmental trust fund and financial mechanism for four international environmental conventions—received USD 3.13 billion in August 2006 from 32 donor governments for its operations between 2006 and 2010.

Despite these impressive figures, the CDM has not elicited the happiness or pride that one would expect. Instead, it stands in a withering crossfire of criticism. Some complain it funds business-as-usual projects, failing to create real emission reductions. Others assail its governance practices, or claim that its projects are too small to incentivize the more substantive emission reductions needed to shift economies toward a low-carbon development path. It is simultaneously too small and too ambitious, and it targets the wrong emission reductions or does not deliver them at all.

The extent of its success may have contributed to these troubles. The EB and independent verifiers cannot cope with the volume of technically detailed work generated by the flood of projects, and industrialized countries fear that more offsets are produced than their emission trading schemes can absorb, lowering their domestic GHG abatement efforts.

With less than six months before United Nations Framework Convention on Climate Change (UNFCCC) negotiators convene in Copenhagen to decide on a future climate framework, it is time to evaluate which of the criticisms are valid and which are expressions of general discontent with the Kyoto Protocol or the concept of offsetting. In this brief paper, I assess whether the CDM has met the objectives in Article 12 of the Kyoto Protocol and compare its performance with the expectations about the role of the mechanism and what it can deliver. I conclude with a short proposal of the mechanism's role in a post-2012 climate framework, and I present a reform agenda to achieve it.

Evaluation of Performance

The CDM's purpose according to Article 12.2 of the Kyoto Protocol is twofold:

- To assist Parties not included in Annex I to achieve sustainable development and contribute to the ultimate objective of the Convention
- To assist Annex I Parties compliance with quantified emission cuts and reduction commitments under Article 3 of the Kyoto Protocol

Applying the letter of the Kyoto Protocol, both objectives have been met. First, it is a developing country's prerogative to define whether a CDM project falls within its sustainable development strategy when it approves the project. Sustainable development is not defined by the Kyoto Protocol or the decisions of the Meeting of the Parties, so all 1,671 registered CDM projects with host country approval are assumed to contribute to the country's sustainable development. The Kyoto Protocol simply does not leave any room to second-guess the approvals and underlying policy decisions of CDM host countries.

Second, the CDM contributes to Annex I countries' ability to meet their emission reduction targets. Since 2000, public and private entities from industrialized countries have used the CDM to lower the costs of compliance with the targets set by the Kyoto Protocol. Most Western European governments have established CER purchase programs or authorized the World Bank to acquire carbon credits on their behalf, and the EU private sector has poured money into the CDM to reduce the costs of compliance with the European Union Emissions Trading System (EU ETS).

Thus, if the CDM has achieved its legally defined objectives, what are the sources of general discontent with the mechanism?

Sources of Unhappiness

A central criticism of the CDM has centered on the nature of sustainable development, and the different understandings of how the CDM can or should contribute to it. Can sustainable development take the form of industrial energy efficiency or landfill gas destruction, or must it be associated with decentralized and small-scale mitigation and renewable energy projects? Does it create unjustified economic rents, or does efficiency in marginal abatement not affect the value of a mitigation action? The most problematic feature of defining sustainable development is that, while the term is widely used, it embodies so many considerations and values that need to be balanced (social, economic, environmental, and ethical) that its substance is often hard to pin down.

As a market mechanism, the CDM searches for the cheapest emission reductions, and it has been more effective in reducing mitigation costs than in contributing more broadly to sustainability. Yet, from a climate change perspective, it is arguably more worrisome that the CDM has not moved developing countries toward sustainable low-carbon development paths. Critics have challenged the prerogative of the host country to define sustainable development and have expressed concern over CDM funds going to projects with little sustainable development benefits (e.g., destroying industrial gases).

A second significant issue is the CDM's climate change integrity. This mechanism's success is dependent upon real, measurable mitigation of GHG emissions. It is crucial that reductions are additional to what would have occurred otherwise. The EB's interpretation of additionality has been debated vigorously. Some authors claim that many registered projects would have occurred in the absence of CDM certification and award of CERs, while others complain that the EB is excessively stringent in its assessment of additionality. The EB's additionality test embodies a counterfactual that can never be conclusively proven. As long as the CDM evaluates additionality through a test that is coupled with a motivation criterion (why did you engage in the project, and did the CDM influence your investment decision?), it is unlikely that a satisfactory solution to these problems will be found. Critics will continue to question the assertions of project developers that CERs are essential, project developers will have trouble accepting a test which contradicts their entrepreneurial spirit (requiring them to explain why the project will fail without CERs), verifiers mistrust project developers and the EB mistrusts verifiers, and academics will continue to find plenty of reason to challenge the whole system.

To add to these complaints, the CDM does not work efficiently. The approval process is ineffective, slow, and guided by political considerations rather than factual competence. The mechanism has failed to develop a regulatory due process to guarantee fundamental fairness, justice, and respect for property rights. The credibility of the CER market depends largely on the robustness of its regulatory framework and the private sector's confidence in the opportunities provided by the mechanism. This confidence is at risk in the face of mounting complaints about the continued lack of transparency and predictability in the EB's decisionmaking. The governance structure should be reviewed and reformed, taking into account the need to provide private-sector participants (not represented in the Conference of the Parties (COP)/Meeting of the Parties (MOP)) with due process and to ensure the conditions for fair and predictable decisions.

Finally, the CDM has yet to produce the requisite scale of emission reductions. To date, incentives have been too weak to foster the economic transformations necessary to prevent developing countries from following high-emission development paths. While the CDM has worked where carbon can add new sources of finance to investments in private-sector-driven projects, it has failed to mobilize emission reductions for larger policies and programs, including decentralized sources of emissions such as transport or building emissions.

Reasons to Keep the CDM

The CDM has leveraged more finance into GHG emission-reducing projects in developing countries than any other international mechanism, more than its designers ever anticipated. There are other reasons to keep the CDM:

- It enjoys broad support among developing countries. In particular, poorer and smaller countries have established their national CDM authorities only relatively recently and are just starting to engage with the mechanism. There is a risk of losing goodwill and cooperation of developing countries in abolishing a mechanism that enjoys widespread support and while capacity-building to participate in it is still ongoing.

- It is a linchpin of the international carbon market, supporting a community of innovative investors and compliance credit buyers, and providing important lessons for scaled-up carbon trading mechanisms.
- It has been valuable in creating awareness of climate change and capacities to address it among sectors and stakeholders not normally involved in climate policy.
- It remains a useful tool to provide access to project finance for emission reductions in most developing countries, especially those that are poorer or smaller, and for some sectors of emerging economies.

The CDM should therefore not be abandoned without considering the associated political costs. The mechanism certainly needs reform, but should we dismiss it as failed experiment, a corrupt and flawed expression of dysfunctional UN bureaucracy? Or should we engage in a reasonable discussion on a feasible reform agenda and a meaningful future role for the CDM?

The Reform Agenda

The CDM is in urgent need of reform. It needs assistance in creating more ambitious and broader incentives for developing country emission reductions. A second generation of market and non-market mechanisms under the UNFCCC is needed.

CDM reform and expansion should be built on three pillars:

1. The CDM's environmental credibility needs to be strengthened by replacing the EB's additionality test with alternative tools to evaluate emission reductions, including clear criteria, sectoral benchmarks, approved multi-project or sectoral baselines, discount factors, and positive lists for certain project classes or projects in least developed countries. A decision should be taken after the EB or UNFCCC has

commissioned a study on the impact of the various proposals on the supply of emission reductions from particular regions or project classes.

2. If the CDM is to survive beyond Kyoto's first commitment period, its administrative procedure must meet international due process standards. Private economic actor firms will invest time and resources in generating, monitoring, and certifying emissions reductions only if they are assured a reasonable degree of regulatory certainty. The CDM governance will have to be put on the right track for the second commitment period, enhancing the predictability of its decisions and private-sector confidence in the system. Professionalizing the EB is an essential step. Full-time, salaried individuals, selected on the basis of their technical and administrative expertise, with sufficient technically skilled support staff, can give the EB the necessary independence and resources to deal properly and impartially with a growing volume and complexity of work. In addition, a review mechanism of the decisions of the EB should be established. This would give project participants, and other entities with rights and obligations under the CDM, the right to obtain review of EB decisions.
3. Finally, expansion of both the scope and scale of the CDM is vital. As a project-based mechanism, it suffers from inherent barriers in promoting broader policy change, in some instances even creating perverse incentives which delay adoption of much needed environmental regulatory measures that would reduce emissions standards. Therefore the CDM must be supported by more ambitious sectoral and policy crediting mechanisms. In addition, there are a number of steps that can be taken to allow the CDM to benefit rural and poor communities more effectively:

    - Removal of barriers to programmatic CDM projects such as energy efficiency, decentralized electricity, heating and cooking solutions, transport, and agroforestry programs.
    - Removal of limitations on forestry, agricultural, and land-use projects to allow for projects on land deforested after 1990, and expansion of covered activities to include projects that promote sustainable management and restoration of forests, peat, and grasslands.

Conclusions

Cassandra voices predicting the CDM's doom fail to recognize the critical role that project-based offset mechanisms, including the CDM, will play in a future climate regime. They are crucial to expanding the scope of emission mitigation, leveraging private-sector investment, encouraging innovation, broadening global support, and securing a political deal. The CDM remains a valuable tool for incentivizing emission reductions in smaller and low-emitting developing countries, and it should continue in sectors that do not form part of more ambitious GHG reduction efforts in emerging economies. Where projects are implemented in the context of broader GHG accounting programs, existing projects can be converted and follow Joint Implementation accounting rules.

However, to continue past 2012 there must be reforms and improvements in its environmental and operational performance. These are essential to counter an alarming tendency among EU and US policymakers to call for the domestic design of international offset mechanisms. Since the demand for carbon credits is mainly generated by emission trading schemes in industrial countries, these countries have the power to dictate the rules of the game. If they decide to wield this power, not only would developing countries lose much of their influence, but the CDM and the CER market could find itself subject to a multitude of conflicting offset standards from Washington and Brussels.

Too much has been learned, and too much remains viable, for policymakers to abandon a functional project offset system. Outlined above are only a few of the reasons why we should extend the CDM's lifeline and why we should all be interested in a robust, credible, harmonized, and universal international offset standard.

FURTHER READINGS

Christiana Figueres and Charlotte Streck, "Enhanced Financial Mechanisms for Post 2012 Mitigation," *World Development Report 2010, Technical Background Paper*, in press.

Axel Michaelowa, Purohit Pallav, *Additionality determination of Indian CDM projects: Can Indian CDM project developers outwit the CDM Executive Board?* (London: Climate Strategies, 2007).

Karen Holm Olsen, "The Clean Development's Contribution to Sustainable Development: A Review of the Literature," 84 *Climatic Change* 1 (2007).

Lambert Schneider, Berlin, Oko Institut, *Is the CDM fulfilling its environmental and sustainable development objective? An evaluation of the CDM and options for improvement* (2007).

Charlotte Streck and Jolene Lin, "Making Markets Work: A Review of CDM Performance and the Need for Reform," *European Journal of International Law* (2008).

Michael Wara, "Measuring the Clean Development Mechanism's Performance and Potential," 55 *UCLA Law Review* 1759 (2008).

C

Sectoral Programs for Emissions Control and Crediting

Chapter 7

Why a Successful Climate Change Agreement Needs Sectoral Elements

Murray Ward

Principal, Global Climate Change Consultancy; Former Senior Climate Change Negotiator, New Zealand Government

Key Points

- Sectoral elements such as sectoral no-lose targets (SNLTs) and other kinds of sectoral agreements are an essential next step to climate change mitigation in developing countries, and can benefit both developing and developed nations.
- SNLTs can benefit developing countries by enhancing the scale of incentives for private-sector investment, motivating sector-wide emissions achievements, and providing linkages to global carbon markets. SNLTs are in effect a type of sectoral NAMA crediting. SNLTs work best for emissions-intensive sectors with a few major sources, e.g., electricity generation, cement, iron and steel, aluminum, oil and gas production, and refining. They typically set emissions-intensity requirements.
- For sectors where SNLTs will not work, other kinds of sectoral agreements can address use of low-carbon technology, technology diffusion, etc.

Sectoral Approaches—New and Necessary

Sectoral elements are a necessary part of any international climate mitigation agreement that seriously engages developing countries. Although there is great variety among sectoral proposals, they all seek to encourage mitigation across a sector of the economy, rather than just on a project-by-project basis. There have been some early missteps along the way regarding what is meant by a sectoral approach, and it is notable that proposals for sectoral elements initially came mainly from developed countries. This has raised suspicions and clouded inclusion of sectoral elements in the global regime for the post-2012 period. But the case for them is strong.

Sectoral—How?

The question is how sectoral elements for mitigating emissions in developing countries might play a role in a global climate change agreement. In practice, it may be seen that the delivery of current mechanisms and programs happens in sectors. In the Clean Development Mechanism (CDM), it is clear that there is significant sector-level specialization. This is true for project developers, technology providers, and those providing both project finance and carbon finance. And in the non-UNFCCC world of the Asia Pacific Partnership and the Major Emitters Forum, most technology cooperation activities have been developed and delivered through sector-specific task groups.

Of particular importance is how these existing mechanisms and programs may be enhanced to scale up mitigation activities in developing countries—and the technology and financing transfers and investment needed for this to happen. Can this be more effectively achieved by taking a sectoral approach? What are the inherent constraints and challenges that need to be addressed? Indeed, the very term "sectoral approach" has been part of the problem in getting these issues discussed in an objective, analytical, and suspicion-free manner. What does it mean—exactly?

What Sectoral Approaches Are Not About

First, one common suspicion has been that developed countries favor sectoral commitments as a means to avoid stringent binding economy-wide emission targets for themselves. But those developed countries that stress domestic sectoral circumstances have increasingly made clear that their objective is simply for these circumstances to gain some recognition as negotiations decide the differentiated level of their economy-wide circumstances. Thus, for example, Japan wants others to understand how efficient its economy is—so it has a high abatement cost curve—and New

Zealand wants others to understand that about 50% of its emissions come from the agriculture sector, where mitigation possibilities for such things as ruminant methane emissions from its livestock are quite limited.

Second, sectoral approaches are not about trying to have industries in developing countries (and their governments) sign up to binding international sectoral agreements. Nor are they about negotiating performance benchmarks for industrial processes that may be the basis for possible border tax adjustments or, in developed countries, the basis for allocations of grandparented allowances in domestic emission trading schemes. For developed countries, then, a *sectoral approach* is not about their emissions (in the international agreement anyway); it is about mitigation in developing countries.

The one exception to this is the special case of international marine and aviation bunker fuels. These emissions arguably should be managed by both developed and developing countries on a sectoral basis. There is an ongoing debate as to whether these should be managed by their respective existing multilateral intergovernmental processes (International Maritime Organization and International Civil Aviation Organization), or brought under a United Nations Framework Convention on Climate Change (UNFCCC) agreement. But this debate is outside the scope of this piece.

A Flexible Approach

So if we now have a better idea of what a sectoral approach is not, do we now know what is? It is described by some as a portfolio of possible measures that can be specific to the sector in question—and also to a given country. For example, in a side event at the June 2009 UNFCCC sessions in Bonn organized by the WBCSD on its Cement Sustainability Initiative, a sectoral approach was described as "a combination of policies and measures, developed to enhance efficient, sector-by-sector, greenhouse gas mitigation, addressing data, policy, technology and capacity building within each sector," with the elaboration:

- International cooperation with major sector actors to develop and share appropriate sector tools, systems, data, best practices, UNFCCC crediting policies, benchmarking, and technology development.
- Nationally appropriate mitigation actions (NAMAs) tuned to a sector. Emission goals and policies could differ depending on national

> ambition, common but differentiated responsibilities, and local circumstances.

Presumably, this rather all-encompassing description of a sectoral approach has resulted from a process in the global cement sector that involves major industry players in both developed and developing countries. Therefore, it may well indicate how the term "sectoral approach" needs to be communicated in a range of sectors to allay suspicions of developing countries. But while this rather broad, all-things-to-all-people ethos is fine, negotiations rightly focus on more specific policy tools that seem to be getting traction in the negotiations.

Sectoral No-Lose Targets

One such policy tool is sectoral no-lose targets (SNLTs). A country will not be penalized if it fails to meet an SNLT, but the country will receive carbon credits if it meets or beats the target. These credits can be sold into the international market for compliance carbon units. There would then need to be some means to translate this national-level incentive to individual investments and changed practices on the ground. This could be through domestic policies, or there are also ideas for complementary international policies.

A related policy tool is NAMA crediting: granting credits for non-binding nationally appropriate mitigation actions that meet certain conditions. If this is done on a sectoral basis in a given country, it is similar to SNLTs. Put another way, SNLTs can be seen as one element of the concept of NAMA crediting.

A key argument in favor of sectoral crediting is that it is not constrained by the additionality-based procedures that have created such complications for both project-based and programmatic CDM. Sectoral targets are by their nature crediting baselines. If these are agreed by a negotiating process (just as developed country targets are agreed), then additionality need not be a concept to be applied (just as it is not when developed countries beat their targets and can sell their surplus units into the market). At the same time, the sectoral targets should be set at levels that avoid crediting for actions that would just happen anyway, including those supported by other financing and technology transfer mechanisms.

SNLTs (or sectoral NAMA crediting) are therefore seen as an enhanced market mechanism that can serve to scale up investment in low-carbon technologies and practices. It also provides developing countries with greater flexibility and domestic control over policies and measures that can lead to credits being awarded.

However, there needs to be a workable metric for the crediting baseline and robust monitoring, reporting, and verification (MRV) systems in place to ensure that it is clear by how much a country has met and beaten its target in that sector. For this reason, SNLTs cannot necessarily be applied to all sectors of all developing countries. SNLTs are likely to be most beneficial for emissions-intensive sectors with a small number of large sources, e.g., electricity generation (and potentially transmission and distribution) and industrial sectors such as cement, iron and steel, aluminum, upstream oil and gas production and refining, etc.

SNLTs are generally proposed to be of an intensity nature, e.g., GHG emissions per unit of production (of cement or electricity, etc).

While for developing countries SNLTs can be a mechanism to scale up local investment, for developed countries they can be a means to help assuage the concerns of domestic constituencies. In order for carbon credits to act as a meaningful incentive for investment in developing nations, developed nations will need to adopt ambitious targets. Such targets are unlikely to be popular with some powerful domestic constituencies under any circumstances, but will undoubtedly be more politically feasible if developing countries also take on more ambitious policies and measures. In this way, sectoral targets can serve as an important transition mechanism between the policies set out in the UNFCCC and the Kyoto Protocol and more ambitious policies essential in the future.

The realpolitik of this is crucial. And in the absence of enhanced market demand for credits, increased supply will just depress the value of carbon to the detriment of the entire market—and slow the technological innovation that is critically needed to move to the next rounds of substantial cuts in emissions.

Using Sectoral Agreements Where SNLTs Are Not Appropriate

A different sectoral approach may need to be adopted in sectors for which SNLTs are not appropriate. For example, SNLTs may not be a practical policy tool for some sectors that may be difficult to monitor or control (e.g., the transport sector, including auto manufacturing or buildings). Also, SNLTs may not yet be appropriate in many developing countries because they first need to develop and implement robust MRV systems.

Sectoral agreements may still have considerable value in these circumstances. Sectoral agreements could include such measures as

- Commitments framed not in emissions terms, but to such things as penetration rates of certain low- and zero-carbon technologies (e.g., percent renewable power, percent carbon capture and storage (CCS) ready coal-fired power plants, vehicle fleet emission intensity standards, new building performance standards, etc.)
- Commitments to technology diffusion through cooperation in technology research and development, technology transfer, joint ventures, intellectual property rights protection, etc.

The general concepts of NAMAs, and the broad understandings of sectoral approach represented by the Bonn cement sector definition, are evolving in ways that accommodate and facilitate just these kinds of actions—for both developing and developed countries.

FURTHER READING

R. Baron, I. Barnsley, and J. Ellis, *Options for Integrating Sectoral Approaches into the UNFCCC* (OECD/IEA, 2008), see COM/ENV/EPOC/IEA/SLT(2008)3.

R. Baron, B. Buchner, and J. Ellis, *Sectoral Approaches and the Carbon Market* (OECD/IEA, 2009), see COM/ENV/EPOC/IEA/SLT(2009)3.

M. Ward et al., *The Role of Sector No-Lose Targets in Scaling Up Finance for Climate Change Mitigation Activities in Developing Countries* (May 2008), available at http://www.GtripleC.co.nz.

Chapter 8

Sectoral Crediting

Getting the Incentives Right for Private Investors

Rubén Kraiem

Partner and Co-chair, Carbon Markets, Climate Change and Clean Technology Practice, Covington & Burling LLP

Key Points

- Sectoral crediting raises obvious concerns for investors in specific projects or activities within a sector, who will be concerned that they may not qualify for offset credits if the overall sectoral target is not met because of forces outside their control.
- One potential solution to this problem is for the host government to indemnify investors for any shortfall in the offset credits awarded to a given project because of failure to achieve sectoral goals due to underperformance by other projects.
- Another potential solution to this problem would be to require countries to submit comprehensive sectoral programs that will specify the contributions of individual projects or activities to the overall target. Once those programs are certified as adequate to meet the overall objective, individual firms could receive credits based on whether or not they fulfilled their portion of the sectoral target, not whether the overall sectoral target was met.

One important and innovative proposal in current climate policy discussions is to abandon the project-based Clean Development Mechanism (CDM) in favor of sectoral targets and crediting, at least for certain carbon-intensive sectors in countries that meet a variety of other criteria. Under this approach, no carbon credits would be issued for individual mitigation projects or activities unless the entire sector managed to meet the sectoral target. This approach has the potential both to scale up mitigation investment in developing countries and to drastically streamline the monitoring and verification process for crediting emissions reductions.

This approach, however, also raises an important concern for prospective investors: why invest in costly mitigation measures if there is a risk that the desired offset credits will not be issued, irrespective of how the individual activity or project performs, because the rest of the sector failed to meet its overall target? This chapter first explains the concept of sectoral crediting and the difficulties that it may present to investors, then outlines a possible solution to the problem that still preserves the central features of the sectoral approach.

Sectoral Crediting: A New Flexibility Mechanism?

Of the three flexibility mechanisms under the Kyoto Protocol, the CDM has had by far the greatest impact. Because of CDM, low-cost abatement technologies have been deployed in important sectors throughout the developing world. Local capacity has been created, and infrastructure put in place for measurement, monitoring, and verification of emission reductions. And CDM has provided an invaluable price signal for carbon abatement. But it has had some important limitations. Qualifying and registering individual projects have been unduly cumbersome, with higher-than-expected transaction costs. The scale of deployment has been small by comparison with the actual abatement challenge. And, most importantly, the overall trajectory of emissions in key industrial sectors throughout the developing world has continued to point relentlessly upward.

Sector-based crediting is increasingly seen as the next-generation complement or successor to CDM. Instead of crediting reductions in emissions achieved by project-level activities, the idea is to credit reductions based on the performance of an entire industrial sector in a given country. Reductions achieved in any one installation or project within a sector will be credited only if and to the extent that sectoral performance reflects an improvement against a baseline or achieves a target set for the sector as a whole.

Sectors eligible for crediting might include power generation or cement and steel production, among others. The performance of the sector would be measured against a sectoral baseline (such as a set emissions level below business-as-usual (BAU)) or agreed target. A sectoral baseline or target could be set by reference to absolute emissions from the sector (i.e., absolute emissions relative to a baseline set below BAU for the sector) or, more likely, on the basis of a carbon intensity target or a level of emissions performance based on a particular technology. The targets can be no-lose targets: credits are awarded if the target is met, but there is no obligation to achieve it or any sanction if there is a shortfall. What is critical is that there is an appreciable course correction on a broad sectoral basis—from a BAU scenario that is highly dependent on carbon-intensive industrial processes to a low-carbon pathway for continued growth. The purpose of sector-based crediting is to provide the necessary financial supports for this effort. The question is, will it work? In particular, is it realistic to expect that private capital will flow to activities that are aimed at generating these sector-based credits?

Risks to Investors Presented by Sectoral Crediting

The challenge, from an investor's perspective, is simple. Most proposals suggest that sectoral crediting can only be accomplished in one of two ways: either (i) the host country is awarded the international offset credits (ex post, presumably) and then allocates them to activities that are deemed to have contributed to reaching the sectoral target, or (ii) the participants in those activities can directly obtain the offset credits, but only if and to the extent that the sectoral targets have indeed been reached. In either case, the obvious risk is that an individual project participant will perform precisely as intended, but that the sectoral target will not have been reached because other entities within the sector have under-performed. This risk could very well discourage both foreign and domestic private capital. Why would anyone invest in generating these credits when there is a crucial element that is, virtually by definition, outside of the control of the investor?

An Alternative Approach

There are several possible answers to this problem, but to accept any one of them will require some adjustment to the assumptions that have thus far informed the international and domestic discussion on no-lose targets and sectoral crediting. The most obvious possibility is that the host country government would assume the risk of other participants' non-performance. In other words, participants who did perform would be entitled to make an indemnity claim against the government for the value of the credits they would otherwise have received. The government would then either fine or take other enforcement action against the under-performers (effectively making the proposed sectoral target obligatory for domestic purposes), or find some alternative source of revenue to pay the required indemnity. In either case, such a solution goes against the grain of the no-lose concept: i.e., the idea that what is involved here is only a carrot and not a stick. More importantly, perhaps, it still leaves the private investor at some risk if the host government simply fails to perform on its indemnity. Unless and until there is a guaranty facility of some kind, akin to the World Bank's Multilateral Investment Guaranty Agency, this risk could well be a major disincentive to investment in all but the most financially secure and reliable host countries.

An alternative solution might be as follows: first, national policy mechanisms for achieving the sectoral targets would need to be established. Depending on the framework, these targets and policies could be established by international agreement, the country or organization issuing the credits, or by the host country. Specific policy mechanisms could include incentive structures, such as payments for environmental services, tax incentives, feed-in tariffs, etc. They could also include internally binding measures, such as performance standards for the relevant installations or a sectoral cap-and-trade system. Individual sector participants would then bid in their proposed contributions to a sectoral goal: a utility, for example, might formally undertake to achieve a carbon intensity goal that is equal to or better than the sectoral target. By collecting these bids, the host government would assemble a portfolio of qualified projects that collectively achieve (or over-achieve) the intended result. The plan would then be presented to the agency issuing the corresponding credits (in the case of the United States, that would most likely be the Environmental Protection Agency), which could satisfy itself that the plan itself is feasible, that it is supported by appropriate resources, and that the total contributions do indeed add up to the sectoral target.

If the individual participant then performed at a level equal to or better than its accepted bid, it could claim those offset credits directly from the issuing agency. If not, then it would owe an indemnity obligation to the host government and/or to the issuing agency. If the sectoral target were not reached at the end of whatever is the relevant measuring period (say, 3 to 5 years from the time when bids were initially received), the target for the succeeding period would be ratcheted up (i.e., would be made more stringent) by a corresponding amount, thus providing a disincentive for over-promising. If the target were exceeded, the additional offset credits would be awarded to the host government for discretionary allocation, thus providing an incentive to the government to set realistic targets, and to ensure proper enforcement and implementation of the relevant policies and measures. The essential point is that individual participants would need to make a specific, binding commitment as to their own contribution, but could then invest without having to account for the risk of nonperformance by the government and/or by the other sector participants. At the same time, a sectoral goal would have been set and appropriate incentives would be in place that would drive the achievement of that goal.

In designing an offset crediting system, the perfect must not be the enemy of the good. What matters most is that the system incentivizes and mobilizes capital, and that the trend and the effect overall be in the direction of a low-carbon path. The above proposal is aimed at accomplishing these goals, while preserving the core advantages of a sectoral approach.

FURTHER READING

Richard Baron, Barbara Buchner, and Jane Ellis, *Sectoral Approaches and the Carbon Market* (OECD/IEA, June 2009), COM/ENV/EPOC/IEA/SLT (2009) 3, 29 May 2009.

Lambert Schneider and Martin Cames, "A framework for a sectoral crediting mechanism in a post-2012 climate regime," Report for the Global Wind Energy Council (Öko-Institut e.V., Institute for Applied Ecology, May 2009).

Chapter 9

Forest and Land Use Programs Must Be Given Financial Credit in Any Climate Change Agreement

Eric C. Bettelheim

Founder, Former Executive Chairman,
Sustainable Forestry Management

Key Points

- Nearly half of the mitigation actions available in the period to 2020 consist of reducing deforestation and improving agricultural practices in the tropics and sub-tropics.
- Developed countries face severe limitations on the cost-effectiveness of mitigation actions they can take by 2020. However, developing countries have significant potential to take cost-effective land use, agriculture, and deforestation mitigation actions quickly.
- A substantial portion of land use, agriculture, and deforestation emissions in developing countries are driven by the struggle of the rural poor to survive. No plan will succeed unless the rural poor are given sufficient financial incentive to abandon those activities in favor of other, less carbon-intensive options.

In light of the increasing understanding of the timing and depth of emissions reductions required to achieve a 2°C target and the relative costs of doing so, the next global climate change agreement will need to create incentives for substantial global mitigation actions to occur by 2020. That timeline is dependent on significant changes in forestry, agriculture, and land use practices in the tropics and sub-tropics. However, these changes will only occur if we create the right incentives for developing countries and their rural poor.

Historical Responsibility and the Need for Immediate Reductions in Developing Countries

The next 10 years are crucial to success in stabilizing atmospheric greenhouse gases (GHG) at a level which offers a real chance of avoiding cataclysmic climate change. Although the industrialized countries are primarily responsible for the urgency of the problem, emissions reductions in developing countries can be achieved much more quickly, inexpensively, and efficiently than reductions in developed countries. In fact, over two-thirds of the most effective and affordable emissions reductions that can be achieved by 2020 must come from opportunities in the developing world. The technology required to achieve deep cuts in emissions from the developed world will simply not be available and disseminated at sufficient scale for another 20-plus years. Current estimates are that only 5 billion of the 17 billion metric tonnes in annual global reductions required by 2020 can be achieved cost-effectively through technological change in the industrial world.

The unavailability of plentiful cost-effective reductions in the developed world challenges the assumptions and dynamics underlying the Kyoto Protocol and the European emissions trading system. Both focus overwhelmingly on forcing dramatic and rapid changes to the energy and industrial infrastructure of the developed world—an approach that was based on a sense of historical responsibility and fairness. Unfortunately, what may have seemed equitable and fitting is neither economically achievable nor environmentally sensible.

While developed countries still must take the lead in reducing their emissions, they must be realistic about the practical limits of what they can contribute domestically by 2020. Over the longer term, to 2050 and beyond, technological change must provide most of the solution. In the meantime, the developed countries must help to enable and pay for far bigger than expected reductions from the rural areas of the developing world.

Benefits of Emissions Reductions in Agriculture, Forestry, and Land Use

A total of 31% of global emissions result from agriculture, forestry, and land use (AFOLU)—17% from deforestation and forest degradation and 14% from agriculture (see Figure 9.1).

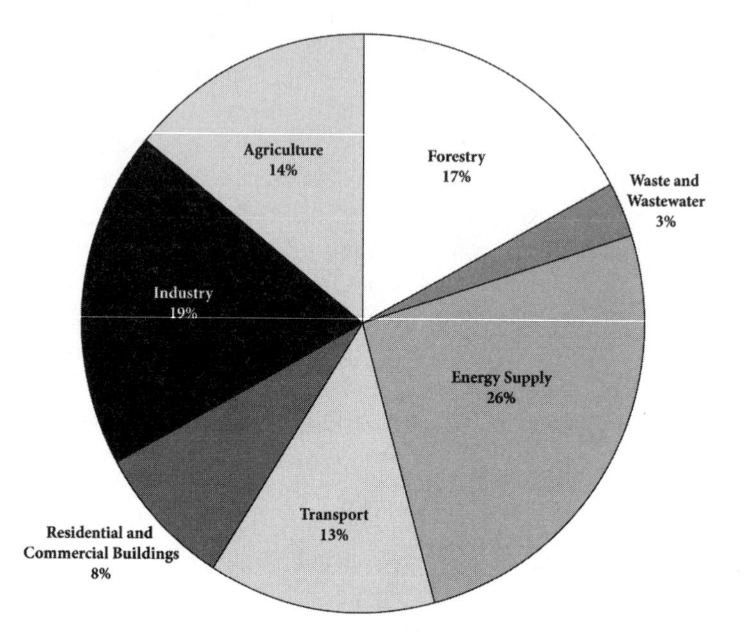

Fig. 9.1. Global GHG emissions by sector (2004). (Source: *Climate Change 2007: Synthesis Report; Contribution of Working Groups I, II, and III to the Fourth Assessment Report of the Intergovernmental Panel on Climate Change,* Figure SPM.3, IPCC, Geneva, Switzerland)

The emissions from AFOLU—90% of which occur in the developing world—offer nearly half, 46%, of the world's potential emissions reductions (see Figure 9.2). This is because photosynthesis is by far the most efficient means of capturing and storing carbon dioxide, and the most cost-effective because it requires neither new infrastructure nor technological breakthroughs. What is required is a change in the economics and regulation of land use. However, to realize these emission reductions, proper incentives will be required—primarily by crediting them in the world's carbon trading systems. If these emissions reductions occur, it buys time for developed countries to put in place greener economies without further depressing global output and competition.

The incentives for crediting AFOLU for industry in the developed world are obvious: low-cost compliance credits in the near term. The incentives for developing countries (forested, deforested, and unforested alike) are even more compelling: significant capital in new investment in their rural areas, higher agricultural productivity and land values, preservation of extant fresh water and biodiversity resources, and poverty alleviation. Further, changes in land use will generate investment capital through the creation of offsets that can fund a country's continued development and transition to a low-carbon economy.

A Market-Based Solution: Changing the Patterns of Land Use

In order to accomplish these goals, biologically stored carbon must become worth more standing up and in the ground than cut down and converted into animals and crops. Fortunately, tropical land which captures carbon, even at relatively low carbon prices, is worth up to 10 times more than the same land harvested for timber and then converted to agriculture. Through crediting AFOLU in global emissions trading, annual payments of between USD 40 and 100 billion to developing countries for the biological storage of carbon could be made. Such funding, together with multi-lateral capacity and institution building programs, is necessary to

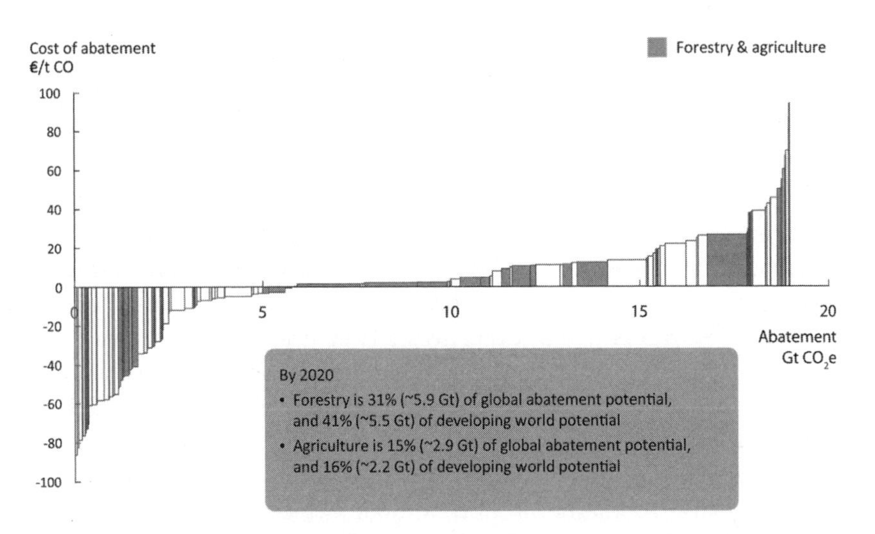

Fig. 9.2. Forestry and agriculture account for 46% of potential global abatement. Developing country abatement cost curve, 2020 (up to costs of €60/t). (Source: McKinsey Global GHG Abatement Cost Curve v2.0 (2009))

counter the current incentives to cut trees and continue unsustainable forest and agricultural practices.

Simply providing capital through AFOLU crediting will not realize this enormous opportunity unless the economics of rural land use is understood. All too often the debate is about the drivers of deforestation as if they were remote from human needs or somehow avoidable—things that could simply be switched off. Nothing could be further from the truth. The real drivers are the necessities of life: food, shelter, energy, and water. 80% of land use change in the developing world is for food, 48% for subsistence farming, and 32% for commercial agriculture. Global demand for food will increase substantially as the population grows from 6.5 billion to a projected 9.5 billion in the next 40 years. Almost all of the population growth this century and the consequent strain on land will occur in the developing world. The land on the planet available per person will shrink from over 5 hectares in 1950 to less than 2 hectares in 2050 as demands for food and higher standards of living increase. The intensification of agriculture is therefore essential if any significant tropical forests are to remain intact by mid-century. That intensification is only possible if significant new capital investment is forthcoming, and if we maintain forests as watersheds and sources of rainfall.

It is also usually overlooked that over 80% of the world's wood harvest comes from native forests. Demand for wood for building material, paper, and timber products is unlikely to abate given population and economic growth. New plantations on a massive scale are required to create a sustainable substitute supply, and this also requires significant capital investment. In addition, fully half of the world's forest harvest is used as fuel for the rural poor. In parts of Africa, it supplies 90% of energy. Any Reducing Emissions from Deforestation and Forest Degradation (REDD) policy must reduce the harvest of native forests and therefore also threatens to remove the only source of energy available to the rural poor of the developing world, as well as decrease the land available to them for food production. These people are widely dispersed and therefore need locally received payments to provide them with the financial wherewithal to buy or build alternative sources of energy and to change their land use. They must receive higher payments than they receive now. No policy prescription or top-down solution will work without providing such payments. In short, unless there are significant new market-based incentives for fundamental change in rural land use practices as a whole, no REDD-only policy will succeed.

National and Global Programs

It is often assumed that the goal is a single market and a single price for carbon. However, that is not what is happening or likely to happen for a very long time. What is happening is the emergence of national and regional trading schemes that will have varying regulations. Any global climate change agreement must provide for this reality as well as set common standards for international recognition of forest and land use credits. Although the EU excludes forest and land use credits, programs in Australia, New Zealand, and the US will include them. By 2012, there will also likely be carbon markets in most large economies including China, India, Japan, Brazil, and South Africa, as well as elsewhere. We must work now to ensure that these countries will adopt standards that will allow for international trading of all forest and land use credits. If we fail in this, the logic of mathematics and economics demonstrate that we will fail to deal successfully with climate change.

Conclusion

Climate change is a global problem in need of a global solution, and developing countries hold the key to success. To give humanity the time and the means to move onto a low-carbon growth path, we must provide rewards and support that bring a green revolution of sustainable development and investment to the rural areas of the developing world. To sum up a complex reality: enabling 50% of emissions reductions via tropical and sub-tropical forests and agriculture in the near future will make 80% industrial reductions by mid-century possible; a winning formula for all.

FURTHER READING

Office of Climate Change, UK Government, "Climate Change: Financing Global Forests," *The "Eliasch Review"* (2008).

Charlotte Streck, Sebastian M. Scholz, "The Role of Forests in Global Climate Change: Whence We Come and Where We Go," *International Affairs* 82:5 (2006).

Ian R. Swingland, John Grace, Ghillean T. Prance, and Lindsay S. Saunders, "Capturing Carbon & Conserving Biodiversity: The Market Approach," *Philosophical Transactions of The Royal Society: Mathematical, Physical & Engineering Sciences* (2002).

Chapter 10

Stock-and-Flow Mechanisms to Reduce Land Use, Land Use Change, and Forestry Emissions

A Proposal from Brazil

Israel Klabin

President, Brazilian Foundation for Sustainable Development

Key Points

- The vast majority of Brazil's emissions are generated by deforestation and other changes in land use—problems that were not well addressed by Kyoto generally or the CDM mechanism specifically.
- The global climate governance regime should use a stock-and-flow mechanism to reduce land use, land use change, and forestry (LULUCF) emissions, by providing heavily forested countries with REDD funding (through credits or loans and grants) tied to specific emissions reductions based on historical deforestation rates, and a dividend based on the total amount of forest stock remaining in that country as a proportion of global tropical forest cover.
- This mechanism provides value not just for avoiding emissions through REDD but also for maintaining and reinforcing forest stocks.
- If this mechanism is combined with targets and incentives to reduce deforestation rates rather than just stabilizing them, significant levels of efficient abatement can be achieved.

One major limitation of the 1997 Kyoto Protocol is that it does not do enough to reduce greenhouse gas (GHG) emissions from developing nations. In particular, it does not do enough to create incentives for countries to reduce emissions caused by agriculture, land use, and deforestation. Recognizing this, there is strong support for emissions targets for the major emerging economies, significant finance and technology transfer from Annex I nations, and a stock-and-flow mechanism to create incentives to reduce land use, agriculture, and deforestation emissions.

CO_2 Emissions from Brazil

Roughly 75% of Brazil's CO_2 emissions arise from changes in land use, in particular the conversion of forests to agriculture and cattle ranching. The portion of CO_2 emissions from the use of fossil fuels is relatively low in the country due to the high proportion of renewable energy use (46.4% in 2007).

There is an urgent need for a drastic reduction of the deforestation rate in the Amazon region, requiring the control of several variables such as the demand for products in forested areas. The wood produced by the forest fluctuates over time, thus making monitoring figures unstable and difficult to obtain, but the Brazilian government intends to reduce deforestation in the Amazon region to 5,740 km^2 per year by 2017. This would be an important step forward to control the current disordered occupation of the forest.

The Failure of Kyoto and the CDM Mechanism to Adequately Address Deforestation

The finance mechanisms established by the Kyoto Protocol were unable to reduce or halt the expansion of GHG emissions in Brazil. Financing for land use, land use change, and forestry (LULUCF) projects was practically nonexistent. Within this broad category of projects, the Clean Development Mechanism (CDM) only allows reforestation projects in areas deforested before 1990 and forestation where there had been no previous forest vegetation for at least 50 years. Such restrictions, considered a serious mistake, were discussed extensively at the Bali Conference of the Parties (COP 13), and reconsideration of these issues will be a major component of any future climate change regime.

Brazilian carbon projects for Kyoto, based on energy efficiency and alternative sources of energy, were clearly at a disadvantage, in comparison to the ones from countries with higher emissions, due to Brazil's starting point of a cleaner energy mix and thus the lower emissions baseline of its power system. Many biofuels, reforestation, and power generation projects could not be considered for accreditation due to pre-existing domestic regulation mandating their implementation.

Therefore, there is a need to rethink new options for scaling up the financial resources necessary for forest protection. Any new financial mechanism should be effective, sustainable, predictable, performance-based, and supported by diversified sources. Many recognize a need to combine non-market financial resources and market-based mechanisms to ensure sustainability of actions.

Creating Incentives to Slow Deforestation

The recent United Nation's report on financial flows and investment estimates that an additional annual investment of USD 200–210 billion will be required by 2030 to reduce carbon dioxide equivalent (CO_2e) 25% below 1990 levels. However, the recent economic turmoil will require some downward revision of this amount due to the emissions avoided by reduced industrial production.

The estimated realistic mitigation potential in developing countries is approximately 7,000 Mt CO_2e in 2020. Most of this potential (5,250 Mt CO_2e) is available at a cost of less than USD 25 per Mt CO_2e. This estimate takes into account reductions potentially available through CDM, reducing emissions from deforestation and forest degradation (REDD), and carbon capture and storage (CCS).

Various proposals have been presented to increase the financial resources available for low-carbon projects. The most effective proposal came from the Group of 77 and China, arguing that the level of funding for adaptation and mitigation projects should be based on defined budgetary contributions from developed countries. For instance, 0.5–1.0% of the gross national product (GNP) of Annex I Parties would give a nominal annual level of funding amounting to USD 201–402 billion.

Stock-and-Flow Mechanism: A New REDD Proposal

The Woods Hole Research Institute and IPAM (a Brazilian think-tank focused on the Amazon) have produced a sophisticated proposal, the Stock-

and-Flow Mechanism, to implement REDD. It guarantees payments for emission reductions as well as dividends for the total amount of forest still preserved by each country. Although Brazil's profile makes it particularly relevant to provide adequate funding structures to reward REDD efforts, this proposal is appropriate for any developing country with forest cover. It is effective not only for countries like Brazil with a large stock and moderate deforestation rate but also for countries with medium or small stocks and high or low deforestation rates.

Using historical data on deforestation as a baseline, one can calculate the emissions reductions generated by a lowered deforestation rate, and these are paid for either through market mechanisms like sectoral crediting or through grants and loans. A fixed proportion of this funding is withheld and set aside into a fund that is distributed among countries participating in this mechanism based on their contribution to the total global stock of tropical forest cover.

If a country emits over its baseline, it will not receive any REDD credits and, in addition, it will be penalized with a reduction in its stock dividend, also reduced in proportion to its deforestation rate in excess of its baseline.

This mechanism has several advantages over the classic, simple REDD crediting mechanism. It provides positive incentives to maintain and improve forest stocks (contributing to biodiversity, water resources, and soil protection), and it does not punish countries that have already taken action to halt deforestation through early action, such as Costa Rica. It ensures that reductions are not no-lose, as emitting below business-as-usual (BAU) levels is rewarded and emitting over BAU levels brings increasing penalties. It also provides incentives for developing nations to put pressure on one another to improve REDD efforts, as each individual country receives more funding if other countries improve carbon stocks. Crucially, this should help combat inter-country leakage while the national baseline combats intra-country leakage.

According to IPAM and the Woods Hole Research Center, the mechanism can be enhanced by including emission reduction targets instead of just defining a baseline using historical deforestation rates. IPAM estimates indicate that a stock-and-flow mechanism with these reduced deforestation targets is the most effective (in terms of mitigation) and second most efficient (in terms of effective CO_2 reductions vs. credits generated) instrument to reduce emissions from deforestation: see Table 10.1.

TABLE 10.1
Comparison of Effectiveness, Efficiency, and Equity for Different REDD Proposals (Target Set at 20% and Withholding Level = 0.75)

Different REDD Proposals	Reduction in Emissions	Efficiency (Effective CO_2 Reductions vs. Credits)
National historical	61%	71%
Higher than historical for low deforestation	66%	69%
Weighted average of national and global	63%	83%
Uniform fraction of quantified stock	64%	57%
Standard stock-flow	65%	99%
Stock-flow with targets (75-20)	74%	89%

Conclusion

Brazil and other developing countries committed themselves under the Kyoto Protocol to reduce emissions in the Kyoto commitment period 2008–2012, but their obligations were not quantified. If the current trend remains unaltered, the contribution of developing country emissions to total GHG stocks in the atmosphere should grow from around 20% of the world total in 2000 to 45% by 2030.

It is our belief that the highest-emitting developing countries (including China, India, and Brazil) should be bound by commitments for their emission reductions, but the least developed countries should not. Ideally, these initial commitments would last from 2020 to 2050. The European Union's (EU) potentially acceptable proposal to accompany these developing country commitments is to reduce emissions across the EU by at least 20% below 1990 levels by 2020—and even to adopt a 30% target if a satisfactory international agreement takes effect.

Emissions mitigation through LULUCF needs to be adequately dealt with if the global climate regime is to achieve the targets necessary to avoid harmful climate change in a cost-effective manner. One contender to produce efficient and effective results is the stock-and-flow mechanism, which will create the right incentives for heavily forested developing nations to engage with deforestation in a meaningful way.

FURTHER READING

BNDES, Amazon Fund documents and institutional information, available at http://www.bndes.gov.br/english/amazonfund.asp.

Comitê Interministerial sobre Mudança do Clima (Brazil-Federal Government), *Plano Nacional sobre Mudança do Clima* (Brasília, November 2008). Portuguese version.

The Government of Norway, *Reducing Emissions from Deforestation and Forest Degradation (REDD): An Options Assessment Report* (March 2009).

UNFCCC, *Investment and Financial Flows to Address Climate Change: An Update* FCCC/TP/2008/7, 26 November 2008.

Woods Hole Research Center, *How to Distribute REDD Funds across Countries? A Stock-Flow Mechanism* (December 2008).

Woods Hole Research Center, *A Revised Stock-Flow Mechanism to Distribute REDD Incentive Payments across Countries* (January 2009).

D

Leveraging Trading to Maximize Climate Benefits

Chapter 11

Mitigating Climate Change at Manageable Cost

The Catalyst Proposal

Bert Metz

Senior Fellow, European Climate Foundation

Key Points

- Even assuming ambitious GHG reductions by developed countries, large additional reductions in developing country emissions are required in order to limit global warming to 2°C. A total of €65–100 billion annually over the 2010–2020 period is needed to finance these reductions and meet developing countries' adaptation needs.
- International carbon markets similar to the existing CDM could provide an additional €15–20 billion annually, leaving the main contribution of €50–80 billion to public funding. It is unlikely this amount of public funding can be put together under the current economic circumstances.
- Several options exist for regulating the carbon market to get more funding from it, achieve additional reductions, and meet a substantial portion of the shortfall. These include discounting credits awarded, allowing developing countries to sell credits only if they also achieve uncredited reductions, and restricting the award of credits to high mitigation cost sectors.
- A novel and more effective option is establishing an intermediary body (or carbon bank) that would use revenues from credit sales to fund incremental costs of mitigation actions in developing countries, thereby capturing the rent that exists in an unregulated market. That

rent then could be reinvested in additional abatement measures. The carbon bank can be centralized or decentralized; the latter approach might be politically attractive.

Background

Realistic estimates for the funding needed to finance mitigation and adaptation activities in the developing world are in the range of €65–100 billion annually on average over the 2010–2020 period. This takes into account the range of abatement activities with moderate and large positive costs and the barriers to finance that will have to be dismantled or overcome.

Where to Find the Money?

There are, in principle, two sources where the money can be found: public funds and the carbon market. The first question is: what would be a realistic number for the amount that can be obtained from public funds? Increasing of official development assistance (ODA) and transfers of funds generated by CO_2 taxes; revenues from auctioning of domestic emission allowances in developed countries; international auctioning of emission allowances to developed countries; and levies on international aviation and shipping are the most prominent proposals under discussion. All have serious limitations. For instance, an increase in ODA has already been promised by developed countries for assisting developing countries to meet the Millennium Development Goals (MDG). Climate change funds should be additional to MDG funds, but government budgets of most developed countries are under serious pressure. For most other options, international agreement is needed.

The Carbon Market

At present, the carbon market—as we know it from the experience of the Clean Development Mechanism (CDM)—is driven by the demand for offsets in developed countries. Therefore, the volume of the financing through the market depends on the developed country emissions reduction targets. If we assume a 25% on average developed country reduction below 1990 rates by 2020 (the lower end of the 25–40% below 1990 range for developed countries collectively as their equitable share of the effort towards keeping the global temperature increase limited to 2°C above pre-industrial), the Catalyst calculations indicate that €15–20 billion per year out of the total incremental cost of developing country climate finance can be covered through the carbon market. This conclusion is based on the assumption that, as is currently the case under the CDM (ignoring the small adaptation levy), offset credits are sold and bought at a market clearing price, and the buyer receives one tonne of credit for every tonne of offset achieved. Under these assumptions, the total value of market transactions would be much higher than €15–20 billion annually, but much of that value would accrue to project developers or brokers in the form of economic rents (the excess of revenues received over project costs including a normal profit to cover capital costs). This would mean that an amount of €50–80 billion would be required from public sources, which, for the reasons above, may not be unlikely.

Can the Carbon Market Be Reformed and Regulated to Deliver Much More, Reducing the Developing Country Finance Shortfall?

There are in principle several ways to deliver more incremental cost financing out of the carbon market through regulatory measures. The simplest is to depart from 1:1 offset crediting and require a discount: developed countries and their firms that want to use offset credits are obliged to buy, for instance, two tons of offsets for each ton credited. Another approach is to have developing countries accept undertaking some reductions themselves and selling credits if they are able to reduce more than what they promised to do anyway. This means some of the incremental costs are paid for by developing countries themselves (although they may be able to earn rents to cover those costs through credit sales on the additional reductions). A third approach is to restrict the award of credits to offsets from high cost sectors (i.e., the power and industry sectors) so that a higher share of carbon market financing is directed to the sectors where the rents (the excess of revenues received from credit sales over reduction costs) are lower, and more reductions can be achieved with the same amount of market financing.

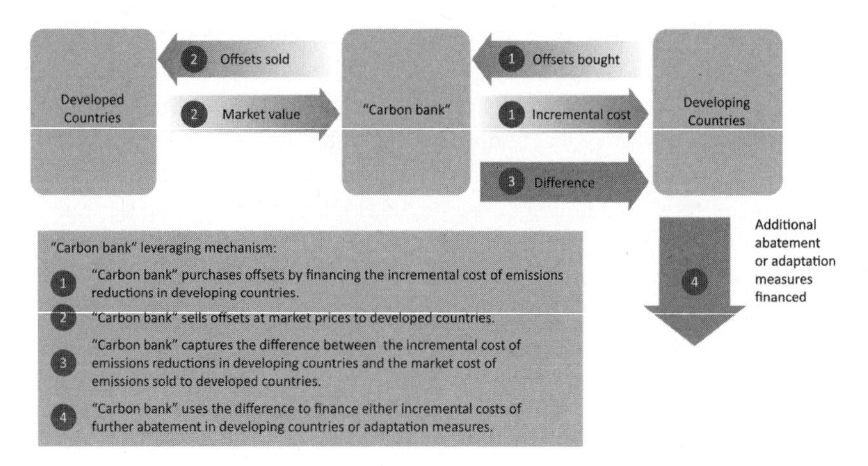

Fig. 11.1. A carbon bank could help to raise additional financing for mitigation or adaptation measures. (Source: McKinsey Global GHG Abatement Cost Curve v2.0 (2009); Project Catalyst analysis)

The fourth approach, one that steps outside the framework that has developed to date, is to create an intermediary body (a carbon bank) that is the sole issuer of credits (see Figure 11.1). This bank would sell credits to developed countries at prices commensurate with the (high) market value of credits in developed countries, but would use the money to cover the incremental costs of the measures in developing countries, eliminating the rents that would otherwise accrue to sellers of offset credits. In principle, the discounting approach can produce a similar additional funding flow, the other two approaches probably less so. In all cases, it is assumed that the least developed countries will continue to have access to a project-based CDM, like what is currently available to all developing countries.

A carbon bank would cover the financing through the carbon market, but in principle could also manage the public funds that have to supplement carbon market financing, creating a basis for integrated and efficient financing.

The carbon bank idea could be implemented in the form of a central international body. This would have obvious advantages in terms of efficiency and transparency, but would not necessarily get the required political support. Developing countries in general are reluctant to accept this centralized model because of their experiences with the World Bank and the Global Environment Facility in disbursing other climate-change-related funds; they complain about lengthy bureaucratic procedures and dominance of World Bank policy over their interests. Developed countries might also have hesitations on a centralized model, particularly if the carbon bank would also handle their bilateral contributions; they like to have control over the destination of their contributions.

However, the bank could also be set up in a decentralized form, either as a series of regional banks or even a network of national banks (maybe regional for small countries). This would enhance the feeling of ownership of developing countries. In fact several developing countries have already set up national trust funds, such as the Brazilian Amazon Fund and the Indonesian Climate Change Trust Fund. For developed countries, a decentralized structure might also be attractive, since it would allow for bilateral arrangements and increase choice.

Where Does That Bring Us?

With a regulated carbon market, the share of the funding coming from the market can likely be increased to €20–40 billion per year. That reduces the pressure on public funding significantly, reducing its contribution to €45–60 billion per year. While it will probably not completely cover the shortfall in funding that can be expected, because €45–60 billion is still a very high number under the current economic circumstances, it provides a much better chance of meeting the required funding needs for an ambitious Copenhagen agreement.

There is also the issue of effectiveness and efficiency of the current carbon market. The project-based CDM is the dominant mechanism in the market at the moment. There are doubts about the integrity of the system, because it is very likely that part of the emission reductions credited through the CDM would have happened anyway; in other words, this leads to higher emissions overall than intended. The carbon market in a post-Kyoto agreement would have to be 5 to 10 times larger than the current CDM. Doing that by scaling up the project-based CDM is not an option. More efficient sector-based program approaches will have to replace the CDM. These program approaches can more easily be controlled to only credit additional action.

The option of the carbon bank, combined with sector-based programmatic approaches has some other advantages over the alternatives: there is a better chance of fixing the current imbalances in CDM financial flows (most money goes to 10 developing countries and many developing countries do not receive anything) and providing funding to all developing countries. A carbon bank would also effectively eliminate the volatility of the carbon price, something that is quite detrimental to investments in low-carbon options in developing countries.

FURTHER READING

McKinsey and Company, *Pathways to a low carbon economy* (2009), available at https://solutions.mckinsey.com/climatedesk/CMS/Default.aspx.

Netherlands Environmental Assessment Agency, "Chair's Summary Report: Where development meets climate: Development related mitigation options for a global climate change agreement," available at http://www.pbl.nl/en/dossiers/Climatechange/Publications/International-Workshop-Where-development-meets-climate.html.

Project Catalyst, *Financing global action on climate change*, available at http://www.project-catalyst.info.

Project Catalyst, *Towards a global climate change agreement—Synthesis Report* (2009), available at http://www.project-catalyst.info/images/publications/synthesis_paper.pdf.

Chapter 12

Engaging Developing Countries by Incentivizing Early Action

Annie Petsonk

International Counsel, Environmental Defense Fund

with Dan Dudek, Alexander Golub, Nathaniel Keohane, James Wang, Gernot Wagner, and Luke Winston

Key Points

- To encourage developing countries to move to low-carbon development paths as swiftly as possible, Environmental Defense Fund's CLEAR proposal (Carbon Limits + Early Actions = Rewards) offers developing countries Clean Investment emissions budgets (CIBs) that can enable developing countries to access a pool of emissions allowances initially greater than their business-as-usual expected emissions, if they place domestically enforceable absolute caps on the emissions of their major emitting sectors.
- By promoting early, broad-scale access to carbon markets, CLEAR seeks to help emerging economies gain access to the capital needed to finance this transition.
- CLEAR provides a measurable, reportable, and verifiable mechanism that rewards any developing country making a firm commitment to reduce emissions early, applying the benefits of carbon trading on a scale far greater than a project-by-project basis.
- CLEAR could also help build capacity early on in a number of areas (technology; abatement opportunities; infrastructure; financial institutions, products, and expertise in the mitigation sector) in developing nations.

Introduction

The world's collective effort to curb climate change will rely heavily upon the global marketplace—the only force large and strong enough to drive the needed innovation and carry through the necessary reductions in greenhouse gases (GHG). This approach is being taken seriously around the world, as evidenced by the success of the European Union Emissions Trading System (EU ETS), the passage of the American Clean Energy Security Act (ACES) through the House of Representatives in June 2009, and proposals under development in a number of industrialized and emerging economies such as Australia, New Zealand, the Republic of Korea, and Mexico. Proposals under discussion in the latter aim to engage in carbon markets much more broadly than avenues currently available through the Kyoto Protocol's Clean Development Mechanism (CDM). Enactment of strong US cap-and-trade climate legislation could, more than any other single step, unite industrialized nations in demonstrating the opportunities presented by low-carbon economic growth.

The effort to prevent the worst effects of global warming will require, however, not just serious emissions cuts by industrialized countries but also early emissions reductions by many others—including, most importantly, the two dozen or so largest, fastest-growing, and most influential emerging economies. This proposal is directed at the this group, offering a framework that can address concerns about limiting emissions without constraining economic growth, and can help generate financing to facilitate the swift and early shift towards low-carbon pathways.

The Basic Idea

CLEAR (Carbon Limits + Early Actions = Rewards) invites developing nations that do not yet have emissions reductions obligations to adopt a Clean Investment Budget (CIB), a multi-year absolute emissions limit covering either the whole economy or the major emitting sectors. Reflecting the negotiations underway in the context of the UN Framework Convention on Climate Change and legislative developments in the United States, nations that undertake nationally appropriate mitigation actions (NAMAs) could propose to the international climate treaty body a Clean Investment Budget (CIB) initially set at levels at or below their anticipated NAMA emissions pathway (Fig. 12.1). Nations could be given access to

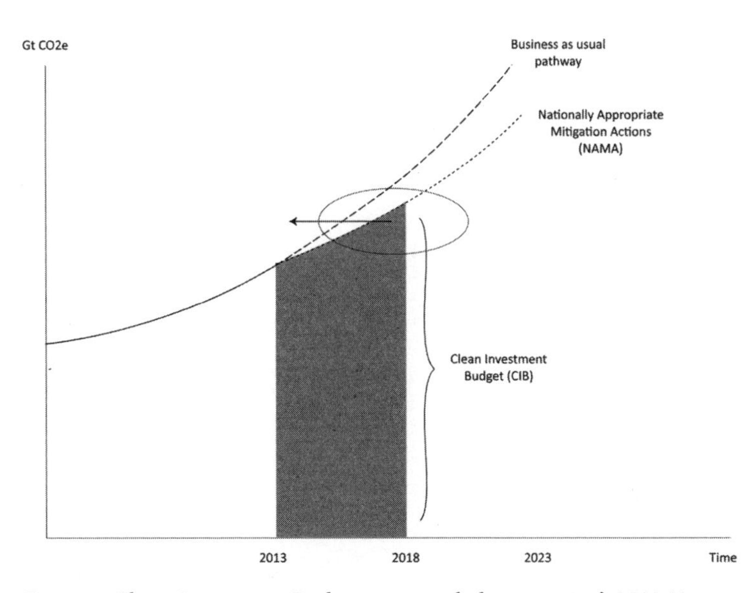

Fig. 12.1. Clean Investment Budget set at or below a nation's NAMA pathway. (Source: Environmental Defense Fund, *The CLEAR path: Rewarding early actions by emerging economies to limit carbon* (2009))

the resulting pool of CIB emissions allowances early, resulting in CIB allowances in excess of the country's emissions at the beginning of the CIB period (Fig. 12.2). These allowances could help provide funding to assist the nations with the transition to a low-carbon economy by allowing developing countries to dock into the carbon market swiftly and efficiently. Many developing countries lack the financing to implement such trajectories. CLEAR taps the power of carbon markets to help nations move swiftly and early to low-carbon pathways.

CIBs would be made transparent, feasible, and enforceable via domestic legislation that binds covered sectors to the declared path. CIBs would need to be determined in advance for at least two successive commitment periods (with the second limit lower than the first), to ensure incentives exist early on to transition to a high-technology, low-carbon economy (Fig. 12.3). Figure 12.3 illustrates a hypothetical CIB over two five-year commitment periods starting in 2013. The upper darker areas indicate the portion of allowances available above current emissions. Initially, only the dark area above current emissions is the area at the beginning of the CIB period. As the CIB delivers financing to help implement NAMAs,

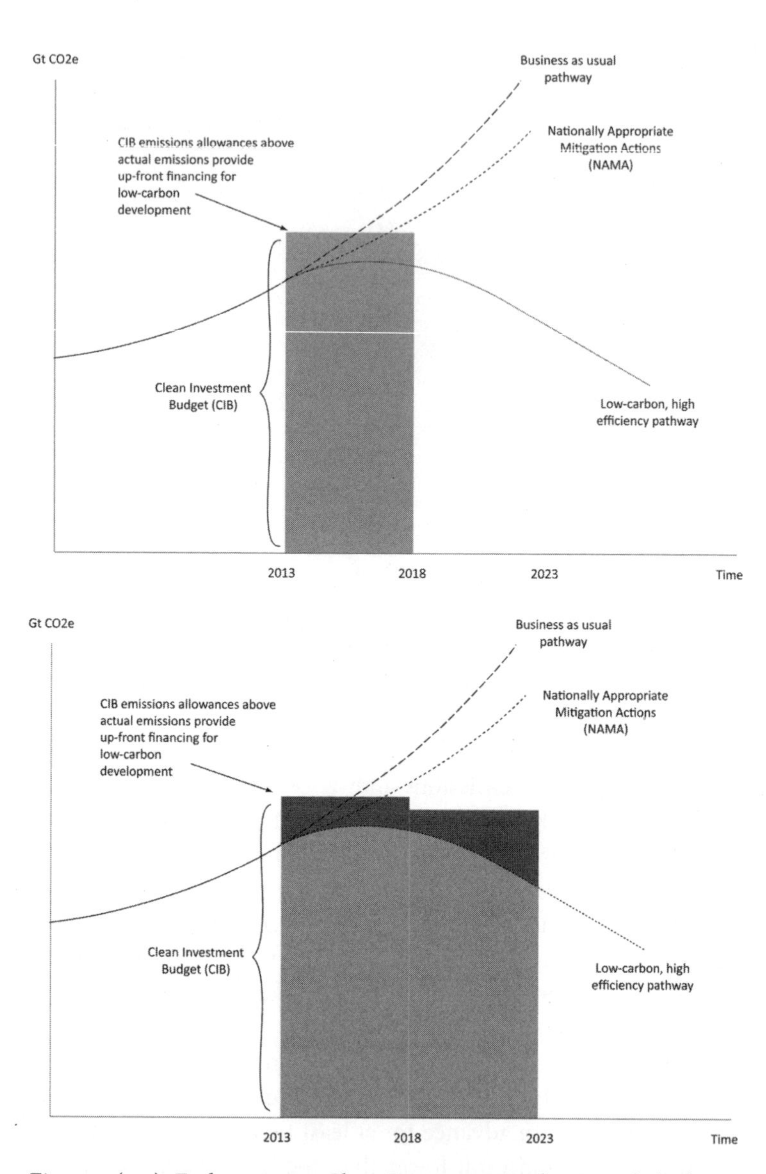

Fig. 12.2 (*top*). Early access to Clean Investment Budgets can help finance low-carbon development. (Source: Environmental Defense Fund, *The CLEAR path: Rewarding early actions by emerging economies to limit carbon* (2009))

Fig. 12.3 (*bottom*). CIB allowances allocated on an average annual basis, for two periods. (Source: Environmental Defense Fund, *The CLEAR Path: Rewarding early actions by emerging economies to limit carbon* (2009)) however, the resulting emission reductions would render more CIB allowances surplus and available for financing more economic and more low-carbon growth, creating a positive cycle for even more ambitious NAMAs, consistent with the Bali Action Plan.

How Much Room Is There for CLEAR?

At the 2009 G8 Summit held in L'Aquila, Italy, and in the Major Economies Forum associated with the Summit, leading nations recognized the importance of averting more than 2°C of warming, a threshold also recognized in the Waxman-Markey climate change legislation that passed the US House of Representatives shortly before the Summit. If, consistent with these developments, nations voluntarily adopt NAMAs that include domestically enforceable multi-year limits on the absolute GHG emissions of their major emitting sectors, set below BAU and at levels consistent with 2°C, they could dock into the carbon market and receive CIBs. Table 12.1 and Figure 12.4 illustrate the constraints implied by a maximum global 2°C increase. Figure 12.4 depicts emissions as indicated in Table 12.1.

Note that Table 12.1 considers the case in which the European Union follows its 20-20-20 approach. If, however, the EU took a tighter target in 2020 of 30% below 1990 levels, and set aside a percentage of its post 2012 allowances to contribute to CLEAR, then at €10–20/ton the tighter EU target could secure a further €24–48 billion in financing from 2013–2020, without any leveraging. Leveraged two to one, it could secure up to €96 billion in financing. While these estimates are contingent upon a number of factors, they are significantly larger than existing flows and rank among the highest proposed new funding mechanisms for GHG emissions

TABLE 12.1
Emissions Targets Assumed to Achieve 2 Degrees (% Difference from 1990 Base Year)

	Country/Group						
	US	OECD Europe	Russia	Canada, Japan, Rest of OECD Pacific	Rest of E. Europe/ Eurasia	Tropcial Deforestation	Other Major-Emitting Developing Countries
2020	–23%	–20%	–10%	10%	–10%	BAU	BAU until 2016;
2050	–77%	–80%	–80%	–80%	–50%	–29%	peak in 2019; then decline

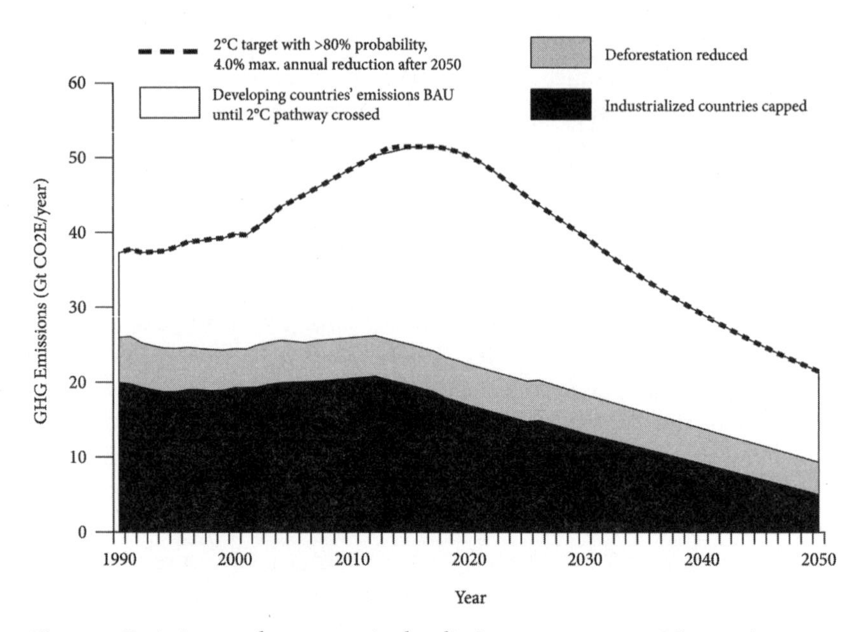

Fig. 12.4. Emissions pathways required to limit temperature to 2°C warming. (Source: Environmental Defense Fund, *The CLEAR Path: Rewarding early actions by emerging economies to limit carbon* (2009))

mitigation in emerging economies. Moreover, CIB allowances could be used to generate even greater levels of financing, as discussed below.

But as Figure 12.4 makes evident, there is little time to spare. Because the availability of CIBs is necessarily contingent upon the gap between existing emission pathways and the point at which the 2°C threshold is exceeded, every year of delay in signing onto a CIB means fewer CIBs that offer sufficient incentives to sign up will be available. The most crucial time for embarking on the CLEAR path is the period between 2010 and 2020, at the latest. If the CLEAR path is not implemented by then and there is no progress toward limits on developing countries' emissions, the atmospheric headroom to accommodate CIBs will disappear by around 2023—even with major emission cuts by industrialized nations. If there is no progress toward emissions limits in emerging economies, progress on emission reductions in industrialized nations is likely to slow, sharply increasing the danger of irreversible, catastrophic consequences from global warming.

The calculation of atmospheric headroom reflects an environmental constraint on the total size of all CIBs available through the CLEAR path. A second consideration is relevant as well: how to make CIBs most effective in the context of global carbon markets.

Achieving Maximum Emissions Reductions through Carbon Markets

A primary goal of the CLEAR path is to provide a readily available source of capital to help emerging economies finance the transition to a low-carbon economy through nationally appropriate mitigation actions (NAMAs). Realizing this goal, however, requires more than simply granting these countries a generous allotment of allowances: a framework must be erected to ensure that CIB funding is well spent. This section sketches out the range of financing mechanisms that could be used.

Financing Mechanisms

One could imagine three broad channels for disbursing CIB funds. First, CIB allowances could be used as collateral to secure traditional financing through private banks or perhaps export credit agencies for emission reduction projects that are "no regrets" or of relatively low marginal cost. The return on investment for these projects would enable the nation to repay the loan and use the CIB allowances as new collateral for a further loan, in effect enabling the CIB allowances to serve as a revolving fund. Used in this way, CIBs would facilitate financing by alleviating the need for alternative loan guarantees and expanding access to credit. Because the financiers would retain their incentive to assess the viability of projects and monitor performance, this approach would require relatively little oversight by the authority holding the CIB allowances other than to perform due diligence on the banks providing the financing, and to ensure that the contract terms were not too generous. Since CIBs might be used only as collateral, a substantial fraction of them would be returned to the carbon capital account after the completion of the underlying loan, and then used as collateral for further loans. Moreover, allowances could be (partially) retired after loan repayment to further strengthen the environmental integrity of the program.

A second option—perhaps less leveraged but also more tightly overseen—could be a system of carbon loan payments or carbon dividends.

In this case, the CIB allowances serve as a guaranteed stream of carbon cash flow. Banks would provide incremental debt or equity financing for emissions reductions projects (in conjunction with other base financing). The host country or project sponsor would repay its debt (or pay out dividends) with CIB allowances. In the meantime, allowances would be held in escrow by the CIB trustee, who would disburse the funds and monitor compliance. The authority would also be responsible for approving the projects and determining their expected yield of emissions reductions. Payments could still be structured to yield carbon leverage of greater than ton-for-ton reductions.

Finally, direct grants, funded by the proceeds from the sale of CIB allowances, would be the most tightly overseen and probably least leveraged alternative. A grant mechanism could be modeled after the Multilateral Fund established by the Montreal Protocol to assist developing countries in reducing ozone-depleting substances, which is commonly seen as a success. As in that case, the responsibility of overseeing national action plans could be assigned to one central, international body, while other entities worked on a local level (the Implementing Agencies in the Multilateral Fund) to approve funding and monitor projects. Grants could be directed at the incremental cost of emissions reductions.

None of these financing mechanisms is sufficient by itself; they are complements rather than substitutes. Using CIB-AAUs as collateral could appeal to countries with well-developed capital markets, and would be suited to projects where an incremental investment is easily identified and yields reliable and significant operating cost savings—for example, energy efficiency in commercial buildings.

Carbon loan payments or dividends would be more appropriate to finance projects where (i) the incremental cost was fairly well defined, (ii) the resulting emissions reductions could be accurately estimated and monitored, but (iii) those emissions reductions fail to translate into financial gains. Finally, grants could be used to finance policies or broader projects (e.g., transmission networks to support renewables) that contribute to long-term reductions in emissions but are less suited to conventional private-sector project finance.

Compliance and Enforcement

Compliance and enforcement are central issues in the design of any international regime; climate policy generally, and the CLEAR path specifically, are no exception. In the context of CIBs, two distinct compliance problems can be identified. First, is the country using its CIB allotment to finance clean investment? Second, is the CIB country meeting its obligation to hold allowances sufficient to cover its emissions?

Each of these problems is individually familiar from international environmental policy. Multilateral development banks as well as private financiers face similar challenges in overseeing how grants and loans are spent in the context of economic development. As in that context, robust oversight of financial flows will be necessary to ensure that countries use their CIBs to fund long-term projects that will reduce GHG emissions in the long run. The stringency of such oversight would presumably vary depending on the financing mechanism used. In particular, when CIB allowances are effectively given to the recipient country as grants, the case for stringent oversight (on both normative and practical grounds) is strongest. When CIB allowances are used as collateral, with the prospect of eventually retiring them rather than releasing them into the market, the potential impact on the atmosphere is much reduced, and thus the need for oversight is as well.

One possibility is for CIB allowances to be held in an escrow account in order to allow for oversight. This, in turn, can serve as a key incentive for compliance, which ought to be especially effective in the early years of the program. If a country has embarked on the CLEAR path and voluntarily taken on a CIB, presumably it will find it valuable in the first few years to comply with the requirements in order to continue to receive the withheld (escrowed) tons. This logic argues for giving large CIBs, but holding most allowances in reserve and releasing them only slowly over time. In this way the CIB can help solve not only the initial participation problem but also the ongoing dynamic participation (continuation) problem. It is also crucial that the escrow account be held as long as possible.

To aid compliance, CLEAR can draw on several risk management tools:

- *Monitoring.* Rules must require reporting of absolute emissions—a crucial element of monitoring-reporting-verification (MRV) without which there is no assurance any climate goal can be achieved. With MRV, market actors are more likely to maintain discipline.
- *Insurance.* CLEAR rules could require nations to offer an insurance pool of pre-agreed allowance quality which could be used as replacements if any marketable CLEAR reductions were challenged.

- *Leverage Limits.* While CLEAR allowances could be used to obtain loans greater than the current value of allowances, rules might establish risk-based leverage limits and require that some CLEAR allowances be held in escrow.
- *Allowance Devaluation.* If MRV indicates that a CLEAR nation is not achieving full emission reduction value, carbon market administrators could devalue its allowances in their trading programs.
- *Plan Robustness.* Plans must be transparent, feasible, and enforceable via domestic legislation that binds the government to the declared path.

Ultimately, as in any agreement among sovereign nations, enforcement cannot be imposed entirely from without. The long-run solution to compliance, therefore, has to rest on ensuring that it remains in the economic self-interest of sovereigns and companies and communities in their major emitting sectors to continue to follow low-carbon development paths. CIBs need to finance investments that make it more attractive ex post to continue along the low-carbon path than to abandon it. Certainly establishing the international and domestic frameworks for such systems will entail overcoming significant domestic political resistance. However, once established, such systems can create an endogenous source of political support, by promoting the growth of clean energy industries with new incentives and resources and delivering clean energy and better living standards to consumers, who then become constituents for remaining in the frameworks. Those domestic constituencies can then help to sustain the political will to continue to participate in low-carbon development frameworks going forward. In effect, the act of participating in the regime helps to reshape incentives in favor of compliance.

Conclusion

A framework that begins with industrialized nations adopting strong binding caps on their absolute emissions, and that invites developing countries to take the CLEAR path, rewards and incentivizes emerging economies to move swiftly to reduce their emissions and increases the chances of avoiding globally dangerous climate change. The sooner emerging economies move to establish CIBs, the greater the rewards they will receive in terms of finance for sustainable development, and the sooner they can transition to more sustainable low-carbon economic development. The greater the delay, the less remaining emissions-absorptive capacity will be available, and the more difficult it will be for the world to avert severe climate change.

FURTHER READING

A. Petsonk, "'Docking Stations': Designing a More Open Legal and Policy Architecture for a Post-2012 Framework to Combat Climate Change," 19 *Duke Journal of International and Comparative Law* 433 (2009), available at http://www.law.duke.edu/journals/djcil/.

G. Wagner, N. Keohane, A. Petsonk, and J. Wang, "Docking into a Global Carbon Market: Clean Investment Budgets to Finance Low-Carbon Economic Development," in *The Economics and Politics of Climate Change* (Helm & Hepburn, eds.), forthcoming, Oxford University Press (2009).

E

Linking Trading Systems

Chapter 13

Carbon Market Design

Beyond the EU Emissions Trading Scheme

Henry Derwent

President, International Emissions Trading Association

Key Points

- The EU ETS has proven the potential of a cap-and-trade scheme to reduce carbon emissions on a large scale. Despite criticisms of the manner of its implementation, its failure to stimulate much investment in low-carbon technology, and its price volatility, important lessons have been learned, and on many fronts it has performed better than expected.
- The concept of offsetting continues to raise political and moral concerns in many quarters, despite supplementarity being taken seriously by developed countries.
- With the future of the CDM currently uncertain, it is important when designing a successor to Kyoto not to lose the benefits of the CDM to developing countries. But it is also important to make sure that new or improved mechanisms are designed in a way that appeals to private-sector investors and—at least initially—covers risks that could put investors off.
- Over time, a global carbon market is most likely to emerge from links established between national and regional schemes. The rate at which this occurs is highly dependent on the balance between the extent of incompatible design features and the benefits to be reaped from such links.

The EU ETS

Much has been written about the European Union Emissions Trading System (EU ETS) and what it has demonstrated about the potential of carbon trading. It is generally acknowledged that the allocation process in the first period, pre-Kyoto, was uncoordinated, and as a result issued too many emissions allowances, giving rise to an embarrassing price collapse. Yet recent econometric analysis suggests that when allowance prices were high, significant carbon reduction did occur. It is also acknowledged that the second phase, though demonstrating the ability of the European Commission to get to grips with excessive national allocations, has been flawed by continuation of conditions making free allocation of emissions permits politically unavoidable, and that the recently concluded framework for the third phase has not improved matters all that much. But over the Kyoto commitment period, the EU ETS has been, as was intended, the premier European emissions reduction policy. It has worked to limit emissions growth.

A further criticism is that even if the ETS has worked in the short term, it has induced little if any investment in low-carbon technologies. Current allowance prices barely justify fuel switching in power stations, let alone the construction of low- or no-carbon generation alternatives. This criticism, however, begs the question of what our objectives should be. The primary economic purpose of emissions trading is to identify and smoke out the lowest-cost emissions reduction options when it is clear that carbon needs to be reduced. There is no justification for installing low-carbon capital equipment quickly if underlying trends in the economy are pulling carbon emissions down, or if there are cheaper untapped reservoirs of low-carbon activity.

Also, the ETS has been criticized for excessive price volatility. Evidence here is usually dominated by the price-collapse in Phase 1. Phase 1 was avowedly experimental and insulated from the Kyoto commitment period. It served to convince the EU member states collectively that the Commission needed to take a tougher role in the future in approving allocations and improving the flow of information. The second period of price weakness, in 2009, is fundamentally different. It would in fact be a matter for some concern if the price of carbon on the EU ETS had not reflected the decrease in emissions caused by decreasing energy costs and production activity due to the economic downturn. This experience in fact shows that the market is working properly. If the Phase 1 story is excluded, the volatility of the carbon price has been no greater than equities for much of the period, and certainly no worse than oil and some other commodities.

However, the argument about the relationship between the EU ETS price and low-carbon investment persists. Critics acknowledge that today's price ought to be no real guide to the expected cost of carbon over decadal time-scales. The real issue is expected carbon prices over the medium to longer run. Here there is definitely a question as to why the EU's commitment to at least a reduction of 20% by 2020, irrespective of the outcome of Copenhagen, has had so little impact on low-carbon investment plans in the EU. Some possible answers are that from now to 2020 is just not a long enough period; or that the Copenhagen international negotiations will provide the final figure, and there is no point in acting in advance of it; or that 20% is in fact a low enough level of ambition to be achievable by a combination of expected regulatory measures and revenue-account activities such as continuing to adjust the fuel mix. There is a persistent suggestion from some firms and academics that longer-term price uncertainty is a political risk that governments ought to underwrite in some way.

Offsets

The degree to which developed country emissions reduction obligations, and the obligations that those countries delegate down to their companies, can be satisfied by any form of offset has been the subject of continuing debate everywhere. On the one hand, achieving a reduction in a global pollutant at the lowest cost available, wherever in the world the reduction takes place, is a central tenet of a rational economic approach to greenhouse gas (GHG) reduction. On the other hand, many people think that developed countries, and companies within those countries, ought not to be given a cheap way out of their obligations and their former profligacy, and ought to be concentrating on reducing their own emissions rather than offsetting them by reductions elsewhere. Plus there are some political objections to paying foreign countries for emissions reductions when there could be jobs or other economic advantage from taking action at home.

This set of arguments was settled in the Marrakech international negotiations by a "supplementarity limit" on offsets, which has been respected in the EU ETS but has still given enough scope for Clean Development Mechanism (CDM) offset credits to enjoy a strong market, and in some cases handsome profits, through EU ETS compliance demand. But the demand for CDM reductions now shows signs of slowing down as many member states have begun pursuing cheaper offset opportunities provided by surplus Assigned Amount Units (AAUs) from other Annex I countries, particularly the generous AAU settlements that Russia, the Ukraine, and other economies in transition took away from Kyoto.

Otherwise the market has in fact done what markets are supposed to do: found the lowest-hanging and lowest-risk fruit first, even if some of these fruits have not been to everyone's taste. Those developing countries that have benefited, and those who think they will have a chance of doing so in the future, have quickly warmed to the CDM even if at an earlier stage their negotiators were hard to convince of its merits. The CDM has achieved investment in developing countries that would not have happened otherwise, some of it involving technology transfer and capacity building; it has awakened interest in emissions reductions in countries that still strongly deny that they have any obligation to reduce; and it has promoted international partnership—all just as it was intended to.

Beyond the CDM Approach

Yet the future of the CDM is now seriously uncertain. The EU has realized that the emissions reduction streams created by the tranches of CDM projects accepted for Phase 2 of the EU ETS will continue to provide a large proportion of the reductions available within the space defined by their supplementarity policy: there is little room left for new offset projects. They have also become concerned that a perpetuation of the CDM will create incentives for developing countries to refrain from adopting emissions reduction targets of their own. And they note that the rise in developing country emissions continues to be so steep that the principle of offsetting developed country emissions with some developing country ones can only have a limited life.

This news, combined with the general uncertainty about global carbon prices, has come close to knocking the bottom out of the CDM market. There are a small number of mainly public-sector funds that are prepared to take a bet on post-2012 CDM 1 prices, but overall the preference is to wait and see. A lot of confidence and capacity is leaking out of the market as a result. It is possible that CDM demand from the US and Australia could come in just at the right time to take over from the shrinking EU appetite, but it is at least as likely that there will be a gap and/or a delay. While Australia has moved in the direction of greater willingness to use emissions reductions available on the world market, there is deep suspicion in the US about the CDM, and much of the available space there for international offsets could be occupied by favored, if still rather vague, forestry offset proposals.

The EU has been setting out a more detailed version of its vision of the future, which is clearly linked to its Copenhagen negotiating strategy. They have specified that the CDM is expected to continue for Least Developed Countries, and that a version of the many-headed concept of sectoral crediting could act as a means of moving the larger developing countries away from project-dependence towards a future where these countries have emissions trading schemes like the EU's that could link in to a wider global system. But apart from the difficulty of persuading the developing countries, whose suspicion of sectoral targets is deep-rooted, little attention seems to have been given so far to the fundamental need to design a crediting mechanism that will be bankable and will attract private-sector investors. Conventional CDM projects look like familiar project finance models, with extra revenue. But how can the disciplines of project finance be applied to a whole economic sector without massive political risk? Who are the counter-parties, who bears the risk, and where is the collateral?

Other New Mechanisms

The same concern arises with other ideas for new mechanisms that are being repeatedly discussed in the international negotiations and in conversations between stakeholders, including other approaches to sectoral agreements such as the crediting of developing country actions that go beyond Nationally Appropriate Mitigation Actions (NAMAs) and Reducing Emissions from Deforestation and Degradation (REDD). It is far from clear what the overall impact on the balance of supply and demand will be, though in general it seems likely that very substantial additional supply could be created through these routes.

Yet clearly, something must be done to engage the larger developing countries more firmly in the enterprise of reducing their business-as-usual (BAU) emissions beyond what they are likely to regard initially as nationally appropriate; there must be a balance between persuading them for geopolitical reasons to go further and paying for emissions reductions that go further still. Those payments are bound to be a combination of public and private money from developed countries.

Clarification of the size of those payments and the proportion that is likely to come from additional public-sector funds or other sources will be very important in order to establish the basis for a political agreement. But the nature and economic justification of the private-sector contribution must also be defined in terms that will make sense to the potential contributors. Carbon finance started as a supplementary revenue source for projects. There seems to be no reason a project-based approach could not continue alongside or within sectoral-level obligations, or alongside the introduction of cap-and-trade as a domestic means of producing a nationally appropriate level of emissions reductions. Departing from the project approach requires a new look at the nature of the investment and its risks, and the potential for risk-reduction. A time-limited or otherwise diminishing availability of these support mechanisms would build on international precedents, although while available they could overlap or run together rather than present a sudden jump between what has been proved successful and what has no record of accomplishment. A new program of policy risk guarantees from international financial institutions (IFI), perhaps including guaranteed levels of emission reduction purchase, could bridge the gap, either directly or passed through the host country government. Special public-private institutions could perhaps be created to define and drive the sector-wide emissions reduction proposals.

The Global Market

The ultimate vision for carbon finance, in the minds of many stakeholders, is a global market. The attractions in terms of economic efficiency are obvious, and there is no reason a global market cannot be consistent with the principle of common but differentiated responsibilities. But it seems very unlikely that major national and regional authorities will submit to a central scheme, particularly if run by a United Nations agency. It is now generally accepted that a global market will instead emerge over time from the gradual coming together of national and regional schemes. But there will probably be a battle between the economic pressures to harmonize and the political desire to preserve design differences resulting from initial national political requirements. Experience with the incompatible design features, such as different forms of offsets or price controls, may make it easier to compromise at a later date, but not necessarily. A great deal of analysis and discussion regarding the various means of linking and unification, and who might gain or lose, is going to be necessary before a sufficient constituency is likely to be built up for sacrificing difficult political choices already taken in favor of a greater good.

FURTHER READING

R. Baron, B. Buchner, and J. Ellis, *Sectoral Approaches and the Carbon Market* (IEA/OECD, June 2009).

Carbon Trust/Climate Strategies, *Global Carbon Mechanisms—Emerging Lessons and Implications* (March 2009).

EU papers on international climate change policy (January 2009), available at http://ec.europa.eu/environment/climat/future_action.htm.

F

Investor Perspectives

Chapter 14

Incentivizing Private Investment in Climate Change Mitigation

Marcel Brinkman
Associate Principal, McKinsey & Company

Key Points

- Reducing greenhouse gas emissions will require significant levels of investment, both private and public.
- Investment in developed countries offers greater investment security due to efficient capital markets and investment processes not found in developing economies, although the latter present more opportunities due to greater rates of economic growth and infrastructure development.
- Up-front capital investment likely will not be attractive to the private sector unless governments provide sufficient cash flow support. Because only a minority of such investments are inherently financially viable, government-mandated incentives such as carbon pricing, standards, and direct subsidies/feed-in tariffs would be required to generate greater investments in mitigation.
- The private sector could respond to incentives that provide a high degree of regulatory certainty into the future and that effectively counter principal/agent problems.

Leaders in many countries are seeking ways to reduce greenhouse gas (GHG) emissions; ever increasing attention is being focused on how the necessary reductions will be achieved. The challenge is significant; if the proposed cuts are to be achieved, the power sector must find new, clean ways of generating electricity; automobile fleets must be replaced with more fuel efficient or electric alternatives; and old and inefficient buildings must be phased out and replaced with new, energy efficient ones.

The global scientific community asserts that the world needs to reduce its carbon emissions to limit global warming to 2°C above 1990 levels. To achieve this limit, the world's nations must stabilize atmospheric concentrations of carbon dioxide equivalents (CO_2e) at 450 parts per million per volume (ppmv), as compared to approximately 385 ppmv today. This requires limiting global emissions to 44 Gt CO_2e in 2020 and to 35 Gt in 2030—a large reduction from business-as-usual scenarios and lower than today's levels (approximately 46 Gt in 2005).

The investment needed to achieve this reduction is significant and presents challenges for investors. National governments do not have the means to invest the amounts required, especially given current economic conditions. Private capital must play a major role in climate change investments, but will only do so within a stable, favorable regulatory and market framework. This means that a key challenge for governments will be to provide sufficient cash flow support to make up-front capital investment by the private sector attractive. The clear implication of this: to create a lower-risk environment that encourages capital investment, policymakers will likely need to provide income support to mitigation projects via domestic regulation.

Where Is Investment Needed?

The McKinsey Green House Gas Abatement Cost Curve (see Figure 14.1) assesses the technical opportunities to abate CO_2e emissions that cost under €60/tonne in the period to 2020, as shown in the graph. Abatement opportunities examined fall into three categories:

- Energy efficiency (buildings, transport, industry), representing 5 Gt
- Low-carbon energy supply, representing 4 Gt
- Terrestrial carbon (forestry and agriculture), representing 10 Gt

Investment in these sectors would start to turn these opportunities into real reductions. McKinsey estimates that in order to reach a desired 450 ppmv pathway, €350 billion of incremental capital investment is needed between 2010 and 2020, and €595 billion between 2020 and 2030. Sector estimates are shown in Table 14.1.

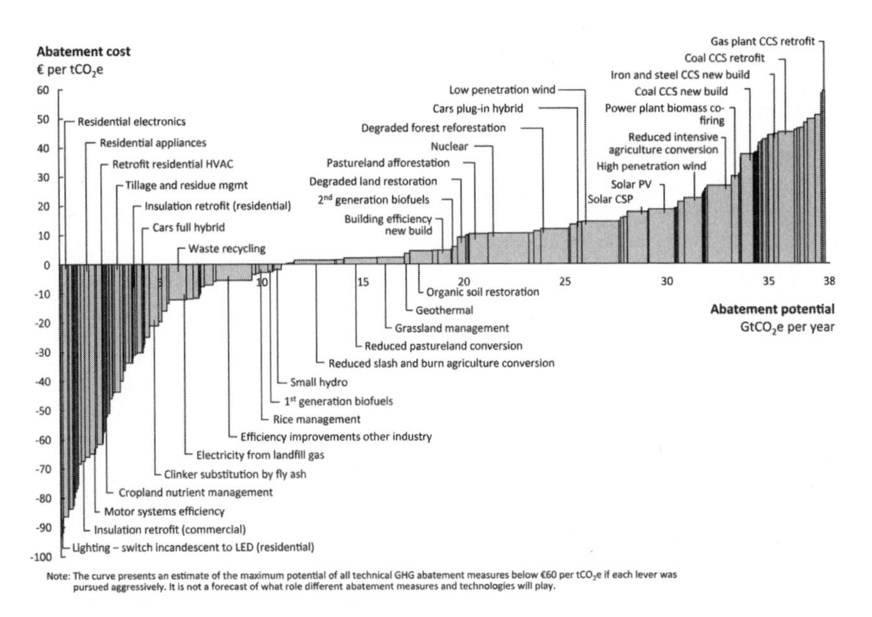

Fig. 14.1. Opportunities to achieve a 450 ppm pathway exist at under €60/t. (Source: McKinsey Global GHG Abatement Cost Curve v2.0 (2009))

TABLE 14.1

Sector	Global Investment Need		Developing Nation Investment Need	
	€ bn in 2010–2020	€ bn in 2020–2030	€ bn in 2010–2020	€ bn in 2020–2030
Buildings (mainly energy efficiency)	€125	€155	€25	€45
Transportation (mainly energy efficiency)	€70	€215	€25	€100
Industry (mainly energy efficiency)	€75	€80	€40	€50
Power	€65	€125	€30	€70
Waste	€10	€10	€5	€5
Forestry and agriculture (terrestrial carbon)*	€5	€5	€5	€5

* Forestry and agriculture (terrestrial carbon) represent a very significant abatement opportunity (10 Gt), but require less up-front capital investment as most of the changes are behavior based, e.g., changed agricultural practices or avoiding deforestation through increased economic activity in and around the forest. The capital expenditure figures shown in the table relate to afforestation, i.e., the investment required to plant trees.

What Are the Differences in Investment Conditions between Developing and Developed Nations?

When considering the investment needed for low-carbon economic growth, the differences in developed and developing nations' investment environments are important. Developed nations have efficient capital markets and investment processes, and should be capable of implementing the right policies to support climate investment. The challenges in developing nations are greater, as investors need to overcome regulatory uncertainty and infrastructure and deployment obstacles. However, the investment opportunities are often also greater due to major infrastructure investments and faster economic growth.

- Developed nations require €220 billion of capital investment per year between 2010 and 2020, and €315 billion between 2020 and 2030: this is mainly driven by replacement or upgrade of existing buildings (47% of the total capital need by 2020) and transportation stock (20% of the total capital need by 2020).
- Developing nations require €130 billion of capital investment between 2010 and 2020, and €280 billion between 2020 and 2030: China represents a large share of this (€60 billion or 44%).

How Can Investment in Mitigation Be Attractive for Countries and for the Private Sector?

Investment requires the right financial and regulatory incentives. Any investment needs to recover the initial investment and the cost of employing its capital over time, adjusted for the underlying risk of the investment. Governments could make the economics of mitigation projects positive for investors; this requires assurances of climate revenues for mitigation via policies and measures that will stay in place, despite changes in government, for the life of the project.

Currently, only a limited number of investments that will produce emissions reductions are inherently financially viable (net present value positive)—those shown on the left-hand side of Figure 14.1. An example would be energy efficiency projects that have energy savings high enough to more than recoup the initial investment. However, even these projects may still need changes in government policies, regulation, or support for up-front financing to realize the potential savings and overcome investment barriers.

Other abatement opportunities require financial incentives to compensate investors for the higher cost of an abatement project relative to alternative investment opportunities. An example might be a wind farm that requires additional financial incentives in order to compete with a high-carbon coal-fired power plant.

Climate and other regulatory policies are the levers left in the hands of government to bridge the gap between returns that an investor requires to make a particular investment and the returns that would otherwise be received. The main methods for incentivizing investors are carbon markets, subsidies, and feed-in tariffs, as well as other policy instruments like standards.

Carbon Pricing

Carbon pricing, either through taxing emissions or through a cap-and-trade or offset credit trading system, affects investment prospects by conferring a monetary benefit on emissions reductions. Attaching prices to carbon through regulation and markets increases the costs of high-carbon technologies and also the market prices of goods and services produced through such technologies, to the benefit of investments in low-carbon technologies. Carbon trading markets also generate commercially valuable carbon credits for low-carbon investments.

Not only does carbon pricing help align private incentives to reduce emissions with public goals, it can also create a revenue stream (either through carbon taxes or auctioning emissions permits) for governments to spend on emissions reduction in other sectors, or to use to reduce other taxation requirements. However, not all emissions can be easily captured in a cap-and-trade market or with a carbon tax, and long-term uncertainty about the level of the carbon price can blunt the incentives provided.

Carbon markets will likely develop rapidly in the next few years, increasing the opportunity for investors. Currently there are two main types of carbon markets:

- Regional and national domestic Emission Trading Systems (ETS), requiring sources to hold emissions permits that can be freely traded. The main example is the European Union ETS; legislation to establish such a system is progressing in the US. Emitting firms are the main actors in this market.

- International (offset) credits generated in developing countries under the Clean Development Mechanism (CDM) can be sold to developed country firms subject to domestic ETS or directly to Annex I countries to meet their Kyoto commitments. Low-carbon projects can earn CDM credits if they prove that they result in emission reduction (i.e., are additional).

The CDM offset credit market has grown rapidly but is still limited in scale (140 Mt of credits generated in 2008) and needs to scale up significantly in order to play a major role in the international financing of abatement. It is questionable whether that will be possible with project-based offsets only. Sector-based schemes, which are typically large scale by their nature, may be required.

Many mitigation technologies are capital intensive and have a long investment horizon, in particular those in the power sector. Relying on carbon markets to provide returns has proven to be problematic in some cases because of uncertainties created by large fluctuations in carbon prices. Many market participants have argued that some form of price regulation or government steps to establish a price floor might be required in order to make carbon market more effective.

Subsidies

Direct subsidies for capital investment or operating expenses, such as those provided by feed-in tariffs in the power sector which reward clean energy with a payment for each kWh generated, promote certainty regarding returns (so long as they are in effect) and have direct positive effect on the investment cash flow profile. Feed-in tariffs have proven to be one of the more effective policies in terms of stimulating investment and have been a policy of choice for many countries. However, they can be expensive for governments unless they are paid for by end users directly.

Regulatory Standards

Mandatory standards to promote climate objectives include engine efficiency standards for automotives and other products, and renewable power standards that require power companies to produce a certain proportion of their electricity from clean sources. Although these standards do not include a direct financial element, they do impose the same standards on an entire industry, thus maintaining a level playing field and passing on costs to consumers through higher prices, thus, in effect, providing an increased return on the investment in abatement. Policymakers like standards as they do not incur costs to the government. As a nonmarket approach, however, they can be inefficient by enforcing abatement even where it is very expensive to do so.

A limited number of best practice regulatory and policy measures can stimulate investment to achieve a significant amount of abatement, often in conjunction with carbon markets:

- Renewable power standards (RPS) can often boost returns from renewable power, making projects viable. Feed-in tariffs are an alternative to RPS; they can act as a guaranteed price for power generated, reducing project risks. Experience shows that feed-in tariffs have been as or more effective than RPS in driving uptake of wind generation.
- Energy efficiency in industry is often linked to upgrading facilities to best practice levels. China in particular is in the process of shutting down many sub-scale production facilities with low efficiency (e.g., in cement) and replacing them with best-in-class facilities, creating opportunities for investors.
- Energy efficiency standards for cars, building codes for houses, and appliance standards can drive innovation and investment in energy efficient technologies and their application. If investors have reasonable assurance that such standards will be maintained and strengthened, they will invest in the likely winners (e.g., car or appliance makers that are already more efficient than the competition and stand to benefit from tightening standards).
- Carbon-content fuel standards open opportunities for biofuels, and make them competitive. Without standards, biofuels are not economic compared to petrol or diesel.

Other Important Elements of Climate Regulatory Policies

When deciding on domestic regulation, policymakers could consider:

- *Regulatory risk.* As discussed above, many climate-related technologies rely on government policies to be economically viable. While some government policies represent credible commitments over longer periods of time (e.g., most feed-in tariffs), others are subject to significant political uncertainty. New Zealand provides a recent example where the planned emission-trading scheme was put on hold after a change in government. Some type of policy guarantee may be required to induce the desired level of investment.
- *Agency problems/industry structure.* Principal-agent problems are a major challenge for energy efficiency projects. In many instances, the logical investors (e.g., owners of apartment buildings in case of building insulation) might not capture the benefits (reduced heating bills) because they will accrue to a third party (tenants). Governments could consider creating alternate business structures like Energy Services Companies, or ESCOs, which invest in (residential building) energy efficiency in return for an annual fee.

Attractive Opportunities for Investment in Climate Change Mitigation Can Only Exist If Current Policies Are Strengthened

To meet abatement targets the world needs €350 billion per year of incremental capital investment in mitigation between now and 2020 in six economic sectors across all nations, developed and developing. Policymakers will likely need to create the conditions that will trigger private investment in mitigation and spur competition among companies to achieve low-carbon economic growth. Well-designed policies, in principle, could spur cost-effective emission reductions, increase energy security, make economies more robust, boost innovation rates, and support economic growth and development.

FURTHER READING

McKinsey & Company, *Pathways to a Low-Carbon Economy: Version 2 of the Global Greenhouse Gas Abatement Cost Curve*, 2009.

Chapter 15

Investment Opportunities and Catalysts

Analysis and Proposals from the Climate Finance Industry on Funding Climate Mitigation

Nick Robins

Head of Climate Change Centre of Excellence, HSBC

Mark Fulton

Global Head of Climate Change Investment Research and Strategy, Deutsche Bank

Key Points

- Investments in low-carbon energy solutions have grown almost five-fold in the past five years. However, investments need to grow a further ten-fold to drive emissions onto a safe path, with almost a twenty-fold expansion in energy efficiency in buildings, transport, and industry.
- Improving regulatory certainty is the lowest-cost option, and this applies at every level from local (e.g., planning permission, fiscal certainty) to international (e.g., policy on technology transfer or credit generation and demand).
- Increasing local capacity to absorb low-carbon finance will also be crucial, requiring the transfer of knowledge from those with expertise in the developed world to their counterparts in the developing world.
- Reducing the perceived risk of investing in low-carbon projects is a crucial step in this process regardless of the success of the previous two options, and the use of credit guarantees backed by public funds and carbon insurance in case of project non-execution or

credit non-delivery will play a key role in making low-carbon investments attractive.

It is clear significant support from private finance will have to be mobilized in order to meet the world's mitigation and adaptation needs in the coming years. As matters currently stand, the right incentives are not in place for this to occur in sufficient volume to have the desired effect. A hospitable climate for low-carbon investment rests on two main pillars: certainty on mid- and long-term targets and a comprehensive policy framework to implement these targets. This paper focuses on the second of these, examining how to both reduce financing barriers and intensify capacity building and knowledge transfer from the developed to the developing world. An overview of barriers to financing is given, before an examination of some key areas where scaled-up investment could have a significant impact such as technology, energy efficiency, and forestry.

Barriers to Financing: An Overview

On the regulatory side, private finance needs long-term regulatory predictability based on transparent rules and procedures at the national, international, and UN levels. Under this regime, climate change institutions such as Designated National Authorities would exist and function efficiently, while markets would internalize the carbon externality. On the financial side, the current difficulty of obtaining debt finance up front for projects, the risk of possible late- or non-execution of the project (including non-delivery of credits), and volatility of carbon prices—assisted by uncertainty on the demand side from cap-and-trade schemes—all contribute to significant project risk that disincentivizes investment.

One approach to overcome these barriers is regulatory in nature, substituting clarity and predictability for uncertainty and opacity in international and national regulation. Another is infrastructure-based, increasing the physical (electricity grid, available resources), institutional, and human (technology workers, public agency capacity, local financial resources, and know-how) capacity to absorb low-carbon investment at the local, regional, and national levels in developing countries.

A third approach is the use of public finance. Debt guarantees backed by public funds, one of the core suggestions of this paper, would significantly reduce project risk caused by any number of the above barriers.

One possibility is the creation of a mechanism whereby the home government of a foreign investor issues guarantees in order to facilitate low-carbon investments in host countries. Examples of these mechanisms currently exist: Overseas Private Investment Corporation (OPIC) and other export credit agencies provide de-risking services, while the Multilateral Investment Guarantee Agency (MIGA), traditionally a guarantor for non-commercial risk, has also been recently experimenting with mitigating commercial risk. Credit risk guarantees and other risk-sharing instruments can considerably lower the investment barriers for many investors and keep the risks associated with direct investments at a reasonable level, even when there exists uncertainty in long-term policy and regulation, local infrastructure, and capacity at the local level.

There is also a desperate need for readily available commercial insurance for low-carbon projects to protect developers and investors across host countries and market environments from risk. One solution is the creation of a Carbon Insurance Vehicle, equipped with public funds but open to private participation. This should be used to insure generation and delivery risks associated with carbon credits, helping to both scale up project activities and assist mitigation efforts in regions and sectors with little activity to date due to perceived risks. This could be centralized or set up at a national level, managed by export credit agencies—possibly integrated with the credit guarantee efforts mentioned above.

The activist fiscal response to the global economic downturn also suggests innovative ways to mobilize capital. HSBC estimates that over USD 500 billion has been allocated to a range of climate change investment themes as part of economic recovery plans. Packages include direct spending, tax breaks, and loan guarantees, with over two-thirds coming from Asia, notably China, Japan, and South Korea. Korea has been particularly assertive in seeing its Green New Deal as a lever for the next phase in its industrial development, deploying public funds to crowd in capital from the national development bank, as well as local pension funds.

The credit crunch has also exposed the inability of capital markets as currently structured to deliver resilient investment returns. Institutional investors are searching for new asset classes and strategies—that can match pension fund liabilities, for example—and the climate economy is emerging as an attractive source of long-term returns. Increasingly climate change is being viewed as another example of systemic risk failure on capital markets, with the failure to adequately price carbon being compounded by incentive-driven short-termism. This continues to result in misallocation of assets to carbon intensive options. Long-term reforms to governance, disclosure, rating, and listing rules are a necessary complement to deliver capital markets that are fit for purpose for the coming climate economy.

Areas Ripe for Greater Levels of Investment

Institutionally the next climate treaty needs to be able to handle a sharp increase in the level of project- and fund-based activity without becoming a bottleneck to climate finance. There should be an increased focus on areas where, to date, the flexible mechanisms under the Convention have had little activity: geographical (Africa and Central Asia), sectoral (energy efficiency, reducing emissions from deforestation and forest degradation (REDD)), or especially in terms of scale (small projects, programs of activities). The last type will require clear standards, reduced procedural complexity, and intensive capacity building at regional and local levels. Again, the use of credit guarantees backed by governments could be of great use in directing finance especially towards programs of activities, as well as the direct deployment of public funds.

Technology

Currently, a number of hurdles to effective investment in low-carbon technology exist. Firstly, there are high transaction costs and timing uncertainties all along the technology innovation process. Secondly, there is a lack of long-term local currency financing options and foreign exchange risks for foreign currency loans, appropriate instruments to manage commercial and political risks, and appropriate intermediaries or incubators to channel appropriate financing and technical support to new entrepreneurs.

Investment in low-carbon technology will require private finance (foreign direct investment), public finance (credit guarantees), and public-private partnerships. Public finance can be usefully deployed to incentivize private finance both in the form of venture capital to bridge the gap between concept and proven technology and project/corporate finance and private equity to fund the deployment of the technology. Developed country governments could provide support through credit enhancement schemes—using their own credit rating to spur low-cost capital flows to private-sector players. Versions of such schemes currently exist in the United States, such as the US Department of Energy Loan Guarantee Program enacted under the Energy Policy Act of 2005, which evaluates renewable energy, carbon capture and storage (CCS), alternative fuel, energy efficiency, and pollution control equipment projects. To date, companies such as Solyndra, Nordic Windpower, and Beacon Power have received USD 594 million in government loan guarantees from this program. In July 2009, the US government announced the most recent round of solicitation by the Program, offering up to USD 30 billion in loan guarantees for various renewable energy projects. In addition, within the American Clean Energy and Security Act of 2009, as passed by the US House of Representatives, there is a proposal to establish a USD 7.5 billion Clean Energy Deployment Administration (CEDA). Under CEDA, a Clean Energy Investment Fund would be established which will provide direct loans, letters of credit, loan guarantees, insurance products, and credit enhancements to support investments in clean energy technologies. In the developing world, additional funding may be necessary. Mechanisms established in the United Nations Framework Convention on Climate Change (UNFCCC) and the Kyoto Protocol, or from the world of development assistance, can serve as models for appropriate funding schemes. Under the convention, the Global Environment Facility and offset mechanisms (i.e., the Clean Development Mechanism (CDM)) are both used to improve project economics for commercially deployable technologies. Using the lessons learned under these regimes, an international financing mechanism for technology development, demonstration, and deployment could be constructed. In development assistance, there are a number of schemes that could serve as attractive models. One is the portfolio of financing options available through the International Finance Corporation. The International Finance Corporation offers a wide range of financing options including loans from its own account, syndicated loans, quasi-equity financing, equity financing, risk management products, and credit guarantees. Another potential model is the USD 6.1 billion Climate Investment Fund of the World Bank. The funds have flexible mandates to co-finance public and private clean technology projects in tandem with other World Bank facilities and other multi-lateral development banks. While the fund is currently more focused on technologies at or near commercial deployment, a similar fund with a mandate for financing earlier stage technologies could have substantial impact.

Energy Efficiency in Buildings

Great mitigation potential exists in commercial and residential buildings, especially through energy efficiency (EE) and greater use of renewable energy. In their construction and occupation, buildings use nearly 40% of the world's energy and are responsible for a similar level of total energy-related CO_2 emissions. Standards should incorporate considerations relating to climate change and sustainability, such as resistance to weather impacts and water efficiency.

Several market barriers prevent these solutions from being effectively deployed: universal limited knowledge of EE opportunities, landlords unwilling to pay for EE measures that lower tenants' utility bills, and tenants unwilling to spend money on property that reverts to landlord at the end of a lease. There are also broader financial/policy concerns: limited access to capital for EE improvements, the need for rapid paybacks, prohibitive permitting requirements, the disparity between project size and large transaction costs, and energy subsidies that discourage conservation.

A number of solutions can be deployed to help overcome these obstacles, in terms of capacity/know-how and finance. The most effective investment may well be in disseminating information to tenants, landlords, investors, and developers about the gains to be had from effective use of EE measures, especially those that more than pay for themselves within a short timeframe. The public sector could have an invaluable role in bringing together financial institutions from the developed world with expertise in EE technology deployment and their counterparts in the developing world, so that this information can be effectively shared. Additional financial incentives will also be needed, in the form of direct subsidies, tax incentives, and credit enhancements; the use of public funds to steer developing country development towards EE deployment in buildings; or using some form of carbon pricing—possibly generation of certified emissions reduction (CERs).

Forestry, REDD

Deforestation of tropical forests has made up 10–35% of global carbon emissions per year since 1990. When standing, tropical forests constitute giant reservoirs of carbon that must be preserved to keep global warming under control. Forests offer the climate change investor the opportunity to sequester carbon and even potentially derive valuable and tradable carbon credits. The key to this is using a sustainable approach to managing the forest and ensuring that the end use of the timber reduces carbon emissions (e.g., second-generation biofuels, housing, furniture). Reforestation of degraded lands would be particularly positive for carbon sequestration. There is clearly a need to include reducing emissions from deforestation and forest degradation (REDD) in the new climate treaty, possibly combining it with national economic development and capacity building programs. Crucially, the private sector should get access to REDD-based carbon credits to properly finance it, bearing in mind the need to address non-permanence in an environmentally credible and financially viable manner.

The American Clean Energy and Security Act of 2009, as passed by the US House of Representatives, would establish a program within the US Department of Agriculture (USDA) to oversee the generation of offset credits from forestry and domestic agricultural sources as a result of a cap-and-trade system. As the current proposal stands, the bill would also require the USDA to establish a set of agricultural, livestock, and forestry carbon sequestration and management practices, policies, and methodologies for project approval and verification measures. In addition, the bill would direct 5% of allowances generated from the cap-and-trade system to secure agreements from developing nations to prevent tropical deforestation.

A wide range of proposals to spur private-sector investment in sustainable forestry are being considered through the UNFCCC negotiations. Ahead of Copenhagen, 17 proposals have been put forward by 25 parties to incorporate forestry more fully into the post-Kyoto climate agreement:

- A coalition including Guyana, the Central African Republic, and others advocates for comprehensive land-based and land-use accounting.
- Australia has proposed that all anthropogenic emissions related to land use, including deforestation, should be included in emissions baselines and in mitigation commitments.
- Indonesia advocates for expanding the sectors covered to include wetland restoration.

- Belarus advocates for adding revegetation, devegetation, forest management, cropland management, grazing land management, wetland restoration, and wetland conservation to activities recognized under the Convention, such as afforestation, reforestation, and deforestation.
- China has submitted a proposal encouraging a tightening up of land-use accounting in developed countries to ensure robust emissions reductions by Annex I parties to the Convention.
- The EU has proposed a series of technical definitions around land use (e.g., "'forest' is a minimum area of land of 0.05–1.0 hectare with tree crown cover (or equivalent stocking level) of more than 10–30 per cent with trees with the potential to reach a minimum height of 2–5 meters at maturity *in situ*.") in order to tighten compliance.

Crafting a well-thought-through regime that unlocks substantial capital flows into sustainable forestry will be critical to achieving low-cost mitigation.

Conclusion

Some general policy points can be distilled from the above discussion. There is a clear need for private finance to be used in a far more significant manner than it has been to date in global efforts to mitigate greenhouse gas emissions, especially in developing nations. Indeed, the UNFCCC finds that 86% of investment and financial flows to address climate change will come from the private sector. The public sector needs to play a role in creating the right incentives for this to occur. These incentives fall into three broad categories: firstly, improving the regulatory uncertainty currently faced by investors at the local, regional, national, and international levels; secondly, creating mechanisms to channel private and public finance to desired locations; and thirdly, using public funds to reduce the risk to which private funds are exposed and creating incentives to divert streams of finance towards low-carbon alternatives. But public funds cannot address the challenge alone. Capital markets themselves need to be modernized to integrate climate risk management into the routine evaluation, allocation, and governance of assets. This is an important agenda for further enquiry that needs to take place alongside national and international climate policy.

FURTHER READING

Deutsche Bank Advisors, Deutsche Bank Group, *Global Climate Change Regulation Policy Developments: July 2008–February 2009* (February 2009), available at http://www.dbcca.com/research.

Deutsche Bank Advisors, Deutsche Bank Group, *Investing in Climate Change 2009: Necessity and Opportunity in Turbulent Times* (October 2008), available at http://www.dbcca.com/research.

HSBC Global Research, *Building a Green Recovery* (May 2009), available at http://www.hsbc.com/1/2/sustainability.

UNEP Finance Initiative Climate Change Working Group, *Financing a Global Deal on Climate Change* (June 2009).

Part III

Bringing Developed and Developing Countries Together in Climate Finance Bargains

Trust, Governance, and Mutual Conditionality

A

Meeting Developing Country Climate Finance Priorities

Chapter 16

Developing Country Concerns about Climate Finance Proposals

Priorities, Trust, and the Credible Donor Problem

Arunabha Ghosh

Oxford-Princeton Global Leaders Fellow, Woodrow Wilson School, Princeton University

Ngaire Woods

Professor of International Political Economy; Director of the Global Economic Governance Programme, University of Oxford

Key Points

- The long history of mutual mistrust between North and South as donors and recipients of development aid is a challenge for climate finance negotiations.
- A stable and secure pool of climate finance is essential. The developed countries must recognize that their promises of future climate financing need to be credible and locked in from volatility or backsliding.
- Trusted institutions for decision-making and disbursement of finance are essential, and the Bretton Woods institutions may not be the answer.
- Effective monitoring, verification, and compliance mechanisms are needed not only for emissions reductions, but also for commitments on financing and technology transfers.

The North-South Negotiating Gap on Climate Finance and Institutions

While there is much variation among different developing and developed countries, overall there is a real North-South gap in climate negotiations. Current proposals on climate financing do not do enough to overcome the lack of trust and mutual credibility between developing and developed countries. This essay analyses the priorities and concerns of developing countries and proposes three planks for a bridge across the gap.

The lack of trust between developed and developing countries reflects not only a lack of appreciation of each other's domestic political commitments and constraints, but also a history of bad faith in the making and implementation of global commitments on development, climate, and institutional reform.

Developing countries view calls to stabilize and reduce their greenhouse gas (GHG) emissions as both illegitimate and a threat. They consider the demands illegitimate because rich countries are primarily responsible for the historical stock of emissions, the atmospheric concentration of which is causing global warming. The calls are a threat because curbing emissions could undermine the growth necessary to lift millions out of poverty. With 1.4 billion people living in extreme poverty, poverty reduction for developing countries is the priority. Any desired action against climate change has to be reconciled to that imperative.

Absent sufficient financing, any commitment to curb GHG emissions would limit the ability of developing countries to increase their energy supply, a central part of their efforts to reduce poverty. About 1.6 billion people in developing countries live without electricity, and 2.5 billion lack access to modern energy sources. Even in fast-growing China and India, more than half of the population relies on traditional biomass for cooking. The easiest and fastest way to increase energy supply under current circumstances often involves the construction of plants with high GHG emissions. However, the provision of financing to improve the efficiency of existing and oncoming modern energy infrastructure will offer a potential win-win situation: many developing countries may be able to use financing from industrialized nations to reduce emissions while increasing economic growth rates. Accomplishing this goal will require substantial technology development, diffusion, and transfer, in addition to financing.

Many developing countries also believe that industrialized nations have not paid sufficient attention to the challenges of adaptation. The imperative for poor countries to adapt (rather than mitigate) strengthens every day as efforts to mitigate by rich countries falter. Yet adaptation is often treated as a side issue, and current spending on adaptation (about USD 1 billion) is a fraction of the estimated requirements.

From a developing country perspective, the demands and priorities of rich countries assume that only their own internal politics matters—and that large developing countries, instead of reducing emissions, are simply stalling. Developed countries seem to insist on setting demanding conditions on recipients of climate financing, and anyhow to be reluctant to provide financing without both setting eventual caps for developing countries and lowering competitive costs for their own economies. Developing countries take the view that since rich countries have repeatedly failed to meet their past commitments on development assistance, any new climate financing proposals will lack credibility unless there is adequate accountability of developed countries to keep not only their emissions commitments, but also their financing commitments to developing countries. Donors have reduced funding or altered conditions even in cases where recipients met specified conditions. Where provided, funding was volatile and unpredictable, thus undermining their best-laid plans. Making commitments for reducing emissions or adopting climate-friendly policies without financial guarantees is, therefore, not politically acceptable to developing countries. Current or potential large emitters among developing countries have some real negotiating power to insist on these demands.

Several developing countries have proposed unilateral climate-friendly measures. China aims to supply 40% of its energy from renewable sources by 2050. Other announcements include Brazil's on reducing deforestation, Mexico's and South Africa's on emissions reduction and stabilization, and India's on energy efficiency and the development of renewable energies. Yet, all are hesitant to sign an agreement that would cap their future emissions without assurances that they will receive substantial technological or financial support from developed countries. They have received little such support thus far.

Developing countries have set out a broad agenda for financing and technology transfer. The G-77 and China have proposed a financial mechanism accountable to the United Nations Framework Convention on Climate Change (UNFCCC) with balanced representation and direct access to demand-driven funding. They also propose a technology mechanism, including a multilateral fund under the UNFCCC, and they call on industrialized nations to divert as much as 1% of their gross national product to help finance emissions-reducing technology projects in the developing world. They want multilateral mechanisms to cover both the full costs (for preparing national communications, patents, and license fees, and for adaptation) and full incremental costs (for mitigation actions, transfer of low-carbon technologies, R&D, and for building institutional frameworks).

In the eyes of developing countries, the approach of developed countries has been inadequate and perhaps even counterproductive. Developed countries seem determined to use the World Bank to channel funding, and even then not the full multilateral International Bank for Reconstruction and Development (IBRD) mechanisms but climate-related trust funds. This is serious because global public financing cannot be avoided. For instance, even with an intervening carbon bank that finances abatement projects in developing countries and sells offsets to developed countries at a profit, 30–45% of annual financing needs would have to be covered using public sources. Moreover, it is unlikely that the sums generated would be sufficient for adaptation activities. Another proposal, which leverages pre-committed emission reduction plans in developing countries to generate loans in the carbon market for mitigation activities, would still raise questions about the predictability of funding in future. In a 2009 submission to the UNFCCC, the United States recognized the need for financing, technology, and capacity-building support but left the section intended for spelling out financing arrangements completely blank.

Holding both sides to account in the financing relationship will be a key element for any financing mechanism to be both politically acceptable and effective. From the above discussion, three principles emerge for climate financing mechanisms and their governance.

1. Ensure the Creation of a Secure Pool of Climate Finance

There must be a credible basis for confidence that a huge gap will not emerge between promised and delivered financial assistance. Developing countries seek guarantees. As the Algerian delegate argued at the Bonn climate meeting in March 2009: "A lottery would not get much ticket sales if it disclosed the prize *after* the draw."

Estimates of the amount of funding required to adequately address global warming vary wildly. A draft report by the UNFCCC's Expert Group on Technology Transfer estimates that additional annual spending on mitigation technologies of USD 262 billion to USD 670 billion would be needed by 2030 (current spending ranges between USD 77 billion and USD 164 billion a year). Due to the variety of estimates, developing countries are hesitant to agree to a set amount of financing, calling instead for industrialized countries to cover the full incremental costs of low-carbon technologies. In light of these requirements, the institutional response so far has been inadequate. Since 1991, the Global Environment Facility (GEF) has allocated only USD 2.5 billion to climate projects and claims to have leveraged another USD 15 billion in co-financing. Its "strategic program," approved in Poznan in December 2008, would devote only USD 50 million to scale-up transfers of technology.

One way to overcome this is to create mechanisms that assure financing without appropriation or interference at the national level in donor countries, such as a carbon tax, aviation, and/or maritime levies, auctions of emission allowances, or direct development assistance. Regardless of the option(s) chosen, the funding would have to be available through a multilateral mechanism to reduce unpredictability and unexpected conditionality of financing.

2. *Use (or Build) Trusted Institutions for Decisionmaking and Disbursement of Finance*

Participation in climate mitigation will not be secured only by financial incentives. Equally vital is the structure of representation in decisionmaking. Many industrialized countries favor the World Bank as a financing and disbursement mechanism. Developing countries have long expressed dissatisfaction with the lack of votes and voice accorded to them in the International Monetary Fund (IMF) and the World Bank, which gives industrialized countries a majority of votes and the United States veto power. Similarly, the Global Environmental Facility (GEF) lacks legitimacy among developing countries because its governance structures give undue weight to the influence of developed countries. Most developing countries have rejected the GEF as a financial mechanism, choosing to treat it only as an operational entity.

The industrialized countries' grip on the IMF and World Bank has led developing countries to exit when they can, in practical terms, from each institution by not borrowing and not taking advice from the institutions (whenever they can afford not to). In climate change governance, exit of this kind could render shared objectives unattainable.

A more specific concern is whether developing countries would have control over the choice of policies and technologies they adopt. The World Bank's Clean Technology Fund (CTF) has been held hostage to US politicians and organizations opposed to financing coal-based technologies, even if the potential efficiency gains and emissions reduction potential for developing countries are large. The CTF's operating mandate requires it to respond to country-owned strategies. Yet, domestic politics in donor countries threatens to close the option of multilateral support for large-scale energy investments in developing countries.

Many of these problems could be resolved through appropriate governance structures that would provide greater representation and control to developing countries. For example, the Adaptation Fund model has managed to avoid replicating the World Bank or GEF representation. Its board, comprising 16 members and 16 alternates, represents the 5 United Nations regional groups (2 from each), the small island developing states (1), the least developed countries (1), Annex I Parties (2), and non–Annex I Parties (2). The CTF also has balanced representation but is hobbled by domestic politics. A new financing and/or technology mechanism would need a similar structure. However, formal seats at the table or voting rights are not enough to secure effective voice and influence. Also important are: the role and selection of senior management; the staffing and location of an organization; the decision-making rules (or form of consensus decision-making); and the capacity of developing countries to identify their own priorities and to hold institutions and their representatives to account. Additionally, proposals to its board should be reviewed by an independent expert committee.

3. Develop Effective Monitoring, Verification, and Compliance Mechanisms for Financing and Technology Transfer Commitments

A third important element of financing mechanisms concerns monitoring and verification. Monitoring financial and technology transfers might be technically easier than measuring emissions. To date, however, reporting on financial contributions has been mixed at best thanks to data deficiencies, multiple sources of funding, and inconsistencies in definitions.

Industrialized countries have emphasized the importance of effectively monitoring emissions. However, developing countries are concerned about the costs of complying with verification systems, as well as potential asymmetries in the application of such systems. In the past, verification and compliance programs have either been ineffective or been applied more harshly against developing countries than against industrialized countries. For example, in the World Trade Organization (WTO), the Trade Policy Review Mechanism (TPRM) has done little to bridge the gaps in information about compliance that would be of use to developing countries. Similarly, the IMF surveillance process is a robust policeman of smaller, poorer developing countries but has little if any effect on wealthy countries. An emissions compliance regime should avoid similar inequities.

A credible financing mechanism in the climate regime would need new institutional features for monitoring and evaluating the effectiveness of financial flows. First, self-reporting by member states should be supplemented by more frequent institutional reporting to measure the origin and destination of financial flows. One option is to use the Organisation for Economic Co-operation and Development's (OECD) Creditor Reporting System. But, if the WTO's new aid-for-trade monitoring mechanism is a precedent, developing countries would demand a dedicated system under the UNFCCC to ensure there was no double counting of assistance provided. Secondly, the data must be also analyzed to evaluate the impact of financial flows. Here the experience of the World Bank and regional development banks in project evaluation could strengthen reviews held within the UNFCCC. Third, knowledge networks could be established at a regional level to facilitate the sharing of information and experience across countries and build capacity for monitoring and evaluation. Finally, compliance-oriented peer review procedures would be needed within the UNFCCC to apply pressure on developed countries to comply with commitments. Discussions about the timeliness, adequacy, and impact of financial transfers should be included in extensive reviews similar to those conducted for emissions and implementation of commitments under Article 8 of the Kyoto Protocol.

FURTHER READING

The ineffectiveness of conditionality is outlined in Tony Killick, "Principals, Agents, and the Failings of Conditionality," *Journal of International Development* 9:4, pp. 483–495 (1998).

For a fuller elaboration of developing countries' experience in other regimes, see Arunabha Ghosh and Ngaire Woods, "Governing Climate Change: Lessons from Other Governance Regimes," in Dieter Helm and Cameron Hepburn (eds.), *The Economics and Politics of Climate Change* (Oxford: Oxford University Press, 2009).

The volatility and lack of predictability in aid financing is examined in Aleš Bulíř and A. Javier Hamann, "Volatility of Development Aid: From the Frying Pan into the Fire?" *IMF Working Paper* WP/06/65 (2006).

On the need for reform in international financial institutions, see Ngaire Woods, *The Globalizers: The IMF, the World Bank, and Their Borrowers* (Ithaca, NY: Cornell University Press, 2006).

On the purpose and politics of monitoring, see Domenico Lombardi and Ngaire Woods, "The Politics of Surveillance," *Review of International Political Economy* (November 2008); and Arunabha Ghosh, "Information Gaps Information Systems, and the WTO's Trade Policy Review Mechanism," *Global Economic Governance Working Paper 2008/40* (Oxford, May 2008).

Chapter 17

Developing Countries and a Proposal for Architecture and Governance of a Reformed UNFCCC Financial Mechanism

Luis Gomez-Echeverri
International Institute for Applied Systems Analysis

Key Points

- There is a pressing need for wholesale reform of the financing arrangements under the UNFCCC at Copenhagen, in terms of scale of funding as well as scope and method of governance structures, with ties to compliance.
- The current situation of financial support for emissions reduction and adaptation support is characterized by a great number of funds with complex administrative processes, minimal transparency or accountability, and conflicting mandates that do not necessarily address or respond to developing country concerns. These funds collectively do not have the capacity to change the course of global development towards a lower-carbon path.
- New efforts to address these issues need to deal with the historical baggage of distrust between rich and poor countries; frame the negotiations within the principle of the legal obligation and compliance to replace the present de facto voluntary system; significantly increase the ability of the UNFCCC to carry out its objectives; bring some measure of harmony and good governance to the multiplicity of funds; and draw significant interest from the private and public sectors, bearing in mind the historical sources of finance for the Convention's purposes.

Introduction

As has been highlighted in a number of chapters in this book, the engagement of developing countries is fundamental to the success of any post-2012 climate regime. Developing country engagement is inextricably linked to issues of governance and institutions. This link is an issue both legal as well as practical. It is a legal issue, because developing country engagement is based on the principle of common but differentiated responsibilities, as well as on the obligations and commitments spelled out in Article 4 of the United Nations Framework Convention on Climate Change (UNFCCC). This Article spells out the commitment of developed countries to support developing country efforts. And it is a practical issue, because without well-designed, well-functioning, and responsive governance institutions—including a financial mechanism that can facilitate and implement mitigation and adaptation funding—the chances of significant developing country engagement are remote. Without this engagement, it will be practically impossible to successfully mitigate climate change.

Flaws in Governance = Flaws in Implementation

Many years ago, developed countries agreed to support the climate change mitigation efforts of developing countries. However, they argued that there was no need for a new financial mechanism, as they believed the Global Environment Facility (GEF), established in 1991, would be adequate. However, it is now obvious to developed and developing countries alike that the scale of funding and the current operational arrangements for implementation are inadequate. The Financial Mechanism—which is meant to play a central role in supporting the implementation of the Convention—is in need of major reform. The reform is required because (a) the severity of the climate change challenge requires a much greater scale of action and response than at present, (b) the need and urgency to act now rather than later to avoid even higher costs and hardship, and (c) the mandates of the Convention.

The inadequacy of the current arrangements of the financial mechanism of the UNFCCC has given rise to a fragmented, complex, and inefficient system of finance for climate change and implementation of the Convention that is characterized as follows:

- A large number of funds and financing instruments have been created to address specific climate-related objectives. Most of these are outside of the Convention, and many of them fund pilot projects rather than large-scale operations.
- Generally, each fund has its own rules of procedure and its own governance structure. Many of them lack transparency and accountability.
- Because of the operational complexity of many of the funds, dedicated experts are required at the national level in order to access and benefit from them. This has major consequences and adds pressure to already weak national monitoring and reporting capacities of developing countries.
- These funds and financing instruments have immense direct and indirect transaction costs.
- The objectives of many of these financing instruments and funds are often formulated neither to respond to the demand or needs of developing countries nor with their participation.
- A majority of these funds and financing instruments prefer to fund projects rather than programs or sector plans of action. This adds to the complexity and transaction costs while at the same time diminishing the relevance and impact vis-à-vis the needs of many developing countries.
- While carbon finance (particularly the offset market) initially had great promise to engage developing countries, it ended up benefiting a small handful. Few projects supported sustainable development or transferred technologies as was initially intended.
- Adaptation—the priority for most developing countries—is vastly underfunded and difficult to attract funding and investment as it cannot be easily integrated into the global carbon finance system.
- As currently designed, the financial architecture neither creates the proper incentives for the transformation toward lower-carbon economies and societies nor facilitates implementation of strategies, plans, programs, and projects for those that do want to take action.

The Challenges

Climate finance negotiations have been seriously affected by the struggle between those who want the public sector to be the major—and even perhaps the sole and centralized—source of funds and those who want the private sector to be the principal vehicle, leaving the public sector to finance only those areas that the private sector cannot adequately fund.

Because of the scale of the effort needed, the solution can only lie somewhere in between. It is impossible to conceive of one public-sector fund that would support all required action on climate change throughout the developing world. Equally important, it is also naive and ill-informed to expect that the current fragmented world with a multiplicity of funds can do the job of supporting the developing countries adequately and, perhaps even more important, that without significant public-sector involvement and support, many of the needed private-sector investments will ever happen, or happen in the areas where they are most needed.

So what should an ideal negotiation seek to achieve? Before even considering the question, the UNFCCC negotiators need to be aware of a few realities that are difficult to ignore and that create a baseline for the negotiations. This baseline and the facts that contribute to it can be characterized as follows:

- One of the principal factors, if not the principal one, contributing to the level of distrust between rich and developing countries has been the issue of finance (the lack of it) and the unhappiness with the present arrangement within the Convention.
- Funding within the Convention has had little relation to issues of compliance to Article 4 of the Convention, a situation that is most likely to change drastically in the post-2012 financial regime of the Convention.
- The level of funding provided has been insignificant when measured against the needs and magnitude of effort needed.
- The current uncoordinated and fragmented world of climate change funding, while often more counterproductive than helpful, is in many ways also a welcome sign of the interest and willingness of many to invest heavily in climate-change-related activities.

A successful negotiation on the finance and implementation aspects of the Convention would therefore need to (a) realistically take into account the historical baggage of distrust between rich and poor countries that now exists and try to address it; (b) frame the negotiations within the principle of the legal obligation and compliance to replace the present de facto voluntary system; (c) create a financial architecture that places top priority to giving the Convention the authority to predictably raise significant revenues to levels that are commensurate with the challenge and to allocate revenues fairly in a way that places the power in the hands of developing countries; (d) create a financial architecture that is given the means to force consistency and harmony amongst the multiplicity of existing funds so as to be part of the overall compliance regime; and (e) create a financial architecture that provides incentives, influence, and guidance for private-sector finance to flow towards climate-friendly investments.

Conclusion: A Framework for Negotiations

As long as the debate on finance and implementation of the Convention remains focused on whether it should be mostly public- or private-sector supported and funded, or centralized versus decentralized, there is little chance that the negotiations will advance in benefit of the Convention. An alternative framework is one where the negotiations would focus on the principal objectives that the financial and implementation architecture would need to achieve, establishing some basic principles that would need to be fulfilled, and matching them to the present realities as the basis for that design.

Funding for climate change today derives from three principal sources and levels. The new financial architecture should build on this reality and adjust it in order to enhance the objectives of the Convention:

- Level I: *The resources that flow through the UNFCCC and that are under the direct authority of the Conference of the Parties* (*COP*). The only financial resources under the authority of the COP are those managed by the GEF, the sole operating-entity of the Convention up to that date. At issue are whether to maintain the present operating-entity system; what the role of the GEF should be in the new regime; and whether all compliance-linked funding would need to flow through or be coordinated by the new operating-entity system. Whatever the decision, the resources under this category—Level I —would be applied directly to the compliance mechanism and to the monitoring, reporting, and verification (MRV) system. It would consist of new funding windows to support areas such as mitigation, adaptation, technology transfer, and capacity building. Level I would be supported by a governance structure under the UNFCCC, with an Executive Board acting as the new operating-entity under the authority of the COP and based on the principle of subsidiarity. As such, it leaves the decision of where to apply the funding (i.e., how to disburse) to countries. The governance structure would need to include national Climate Change Funds and implementation hubs that are linked to the UNFCCC system, the MRV system, and the system of compliance. The institutional structures would vary according to the needs and capacities of countries. But as a minimum, these national entities would need to have the capacity to assess needs and priorities and be in a position to make decisions on disbursement to programs and projects with the most potential for addressing the various thematic area needs. These national entities would also have the responsibility to coordinate and harmonize the disbursement of funding, to promote stakeholder consultation, and to ensure that the climate change programs and projects funded are well-embedded in national development strategies and plans and preferably in nationally prepared climate change strategies or plans.

Some initial target or baseline for this level could be established for 2010 with an agreed rate of increase over a 5-year period. These initial resources would be directed to attend urgent priority needs in adaptation, particularly for countries that are most vulnerable; to fund all National Adaptation Plans of Action (NAPAs); to kick-start an urgent program on reduced emissions from deforestation and degradation (REDD); to support the implementation of Nationally Appropriate Mitigation Actions (NAMAs) as they enter into the mainstream; and lastly, activities in support of technology transfer and capacity building which in turn should become the dedicated and sole areas of responsibility of a new and reformed Global Environment Facility.

- Level II: *The many dedicated public-sector international funds that have been created but that are now outside of the authority or influence of the COP.* It is difficult to imagine that the present chaos of multiplicity of funds would remain unregulated. It is not only ineffective from the point of view of the Convention objectives but also extremely inefficient. Should negotiators, therefore, insist that all of these funds go through the UNFCCC and be under the direct authority of the COP? This would be hardly realistic, particularly since most of these funds are dedicated to creating and strengthening the enabling environments for action on climate change. What is more important is that these funds be placed under the overall oversight of the UNFCCC, which would have the responsibility to provide guidance and assess whether these funds are adhering to the principles established under the agreements reached in Copenhagen. Ideally, a system to link these funds to compliance and the MRV system should be created with the caveat that the principal avenue for compliance-related funding is Level I. Level II should be seen as complementary to the system of compliance, and one that would be several times as large as Level I in the initial stages as countries build their capacities, strategies, and plans of action. With time, this level should decrease with Level I increasing in order to support implementation.

- Level III: *The private-sector and carbon finance that now flows unregulated and often operates with little transparency, oversight, or guidance.* This is by far the largest source of resources for implementation of the Convention. But the full potential of this resource will never materialize unless Levels I and II are organized to break down market barriers; create the enabling environment, policies, and regulations; and increase the capacity of countries to influence and direct these resources to climate-friendly investments and climate change priorities. Levels I and II would concentrate their efforts in leveraging these resources to a scale several times higher than Levels I and II.

A truly effective financial architecture would be one that would mobilize and influence the flow of resources on a scale several times that of what exists today, and that would provide a framework that can enhance consistency, harmonization, and investment promotion in climate-friendly investment at the national level.

FURTHER READING

For more detailed information on the Reformed Financial Mechanism proposal, see Benito Muller and Luis Gomez-Echeverri, *Reform of the UNFCCC Financial Mechanism: Part I: Architecture and Governance* (Oxford: Oxford Institute for Energy Studies, April 2009).

Chapter 18

Climate Change and Development

A Bottom-Up Approach to Mitigation for Developing Countries?

Navroz K. Dubash

Senior Fellow, Centre for Policy Research, New Delhi

Key Points

- In the immediate future, bottom-up approaches, such as NAMAs, for developing countries may have substantial environmental advantages over top-down approaches.
- Top-down approaches based on emission caps risk creating counterproductive incentives, such as incentives to set overly high emissions targets or to avoid early action in order to receive greater financing and higher caps later.
- Top-down approaches may in practice reduce, rather than increase, the predictability of emissions levels and of emissions reductions against BAU baselines or meaningful targets.
- Strengthening domestic institutions in developing countries is needed for successful low-carbon development; strategies to do so are an essential part of a low-carbon development and financing program, but are underemphasized in top-down approaches.

A top-down approach—specifically internationally specified and binding national targets and timetables—has long been the preferred position of environmental advocates. But bottom-up approaches, such as policy measures to be devised on a country-by-country basis, have also been part of the policy grammar of the climate negotiations. In the process of fleshing out the Bali Action Plan, one articulation of a bottom-up approach, nationally appropriate mitigation actions for developing countries, is attracting renewed support. What should we think of such bottom-up proposals?

Background: The Push toward a Top-Down Approach

For those who put climate change mitigation first (as opposed to those who seek to preserve sovereignty or emphasize untrammeled economic growth), a focus on targets and timetables is an article of faith. Indeed this is the best way of ensuring meaningful action from Annex I countries. Many of these advocates also believe that some form of hard targets is the best way of inducing serious mitigation from the developing world. They react with considerable unease to the political support for a formula of top-down caps for the North and bottom-up mitigation actions for the South. They thus welcome proposals to hasten the adoption of some sort of caps for the South, such as incentives for early adoption of caps, offers of no-lose targets, and the like. For example, early adoption of a commitment to reduce emissions below projected emissions could trigger rewards such as eligibility to sell the resultant emissions reductions in a carbon market. Alternatively, some call for embedding developing country mitigation actions in a binding planning framework to enable predictability of action. Under this approach, countries would develop bottom-up measures, but then be asked to aggregate these into a larger national plan, to which they would be held accountable.

While there is no doubt that caps for all would, in theory, be the best environmental outcome, in the current negotiating context, focusing in the short run on explicit caps (or the implicit caps of climate plans) for developing countries is a misguided policy. It will not produce predictability of future emissions from current baselines, and in the short to medium term may be misguided for environmental reasons, quite separate from all the conventional arguments about differentiation, equity, historical responsibility, atmospheric space for economic growth, and the like.

To begin with, what are the arguments for inducing developing countries to take on some form of caps or agree to develop binding plans? The primary argument emerges out of climate science as summarized by the Intergovernmental Panel on Climate Change (IPCC). If, as a global community, we are to restrict temperature rise to between 2 and 2.4°C, we must reduce emissions by at least 50% from their 2000 levels by 2050. Even if the North does take on ambitious absolute caps, additional limitations must be accomplished in the South to achieve a 50% reduction. Taking the next step of converting the de facto cap for the South as a whole to national-level caps, however articulated, is arguably necessary to ensure that the global community is on track toward this global goal. A national-level cap, even if not articulated in terms of absolute emissions, will also send economy-wide signals, and enable integration with global carbon markets. Thus, it is argued, caps or binding plans must be adopted in order to ensure the measurability and predictability required to maximize incentives for mitigation action and to achieve ambitious climate goals.

Both parts of this conclusion are questionable. Under the prevailing conditions of institutions and governance in developing countries, top-down approaches may well not be the best way to incentivize low-carbon development. Moreover, efforts to quantify developing country contributions toward global emission reduction goals may, ironically, discourage the desired early climate mitigation actions and undermine predictability. I refer to this effect as a climate policy uncertainty principle. Below, I expand both these arguments.

Low-Carbon Development Needs Effective Institutions in Developing Countries

To understand the prospects for low-carbon development, we first need a perspective on the process of current and future development in developing countries. Development, I suggest, is not just economic growth from a lower base, but a qualitatively different process than economic growth in industrialized countries. A now substantial literature suggests that successful development is closely tied to the nature of economic, social, and political institutions. By institutions I mean the rules of the game, both explicit and implicit, that guide and shape incentives for decisionmakers. Jump-starting development requires appropriate institutional change, and under-development is to a significant extent a result of persistent and poor institutions.

Under these conditions, top-down measures such as emission caps designed to change relative prices, signal economic opportunity, and stimulate actors to capture efficiency are blunted and can even produce distorting effects. Where existing institutions limit choices or create perverse incentives, inducing institutional change through the political process should be the primary task. Absent this change in the underlying incentives, shifts in relative prices are likely to accomplish little.

The Indian electricity sector provides a good example. For at least a decade, there have been considerable economic gains to be had by reforming the Indian electricity sector, but little has changed. The sector is trapped in a vicious cycle of high loss levels and theft, a growing subsidy burden, and declining service quality. Reversing this cycle would lead to considerable financial, social, and environmental gains through more efficient and equitable electricity production and distribution. But reform of Indian electricity has largely failed due to the interlocking of political interests and governing institutions in the sector. Climate-driven economic incentives would increase the potential economic gains from reform of the electricity sector. They would, however, do little to address the entrenched politics and institutions that block their achievement. Instead, bottom-up institutional reform, backed by clever political dealmaking, is required to change the dynamics in the sector.

The general point is that the more imperfect the institutions, the more markets will be missing or incomplete, and the less useful price signals will be as a driver of change. Bottom-up mitigation actions, forged in the crucible of domestic political debate, are more likely to ensure institutional commitment to carbon reductions goals and perhaps even promote institutional change than are top-down mitigation commitments.

Perverse Incentives Created by a Top-Down Approach: The Climate Policy Uncertainty Principle

The argument so far has suggested that top-down measures, and the price signals they send, are an incomplete and partial solution to climate mitigation. But when we consider their effect in giving countries incentives to game the climate regime—a climate policy uncertainty principle—top-down caps may be downright pernicious. The only form of caps that guarantee the environmental integrity of the climate regime is an absolute limit on emissions. However, absolute caps for developing countries are not on the negotiating table, at least in the short to medium term. All the other forms of caps under discussion introduce serious incentive problems of various sorts.

For example, proposed reductions from a business-as-usual (BAU) trajectory encourage strategic negotiation about what such a trajectory is likely to look like. Given large variations in economic growth rates over the last few decades, there is little basis for an objective definition of BAU. Given this uncertainty, there is a risk that that BAU will be defined generously in the interests of a political solution, leaving considerable scope for developing countries to generate and industrialized countries to buy offsets, benefiting both groups economically but compromising the environmental integrity of the regime.

Indeed, any approach that requires construction of a counterfactual baseline against which to judge progress risks repeating, at a larger scale, the problems of gaming and high transaction cost that have characterized the Clean Development Mechanism (CDM). Recent efforts to develop sectoral approaches carry the promise of lower transaction costs because any such costs are distributed over potentially much larger gains at the sector rather than the project level. However, even here, discussion has been bogged down over whether a sector baseline should exclude measures that are in a country's national interest anyway, either because they bring other co-benefits or can be achieved at negative cost and therefore should not be eligible for any climate-related incentives or support. In practice, putting any sectoral reforms into various buckets—such as negative cost, co-benefits actions, and positive cost—is a negotiation-intensive and potentially counterproductive task. Framed thus, countries have an incentive to demonstrate that as many actions as possible carry positive costs, and to do so by simply not undertaking actions unless they are linked to climate financing. Thus, many discussions over sectoral approaches carry exactly the wrong incentives—they discourage early action and reward stonewalling and late action.

These are only two among many examples of counterproductive incentives created by aspects of a top-down approach. It is altogether possible that the harm caused by such incentives would outweigh the benefits of a top-down approach, or at least reduce its effectiveness to a level below the effectiveness that could be provided by a bottom-up approach.

Answering the Objections to a Bottom-Up Approach

Is a bottom-up approach based on nationally devised actions really a viable alternative to the various top-down approaches under discussion? The attractiveness of this approach lies in the potential alignment of interests between development actions and climate mitigation. Why would a developing country not aggressively pursue developmentally useful mitigation actions that yield climate co-benefits, especially if supported by financing from industrialized nations? Without the threat of imminent caps, developing countries are more likely to aggressively pursue such policies. In the medium to long run, a co-benefits approach may not be sufficient, and developing countries, too, may well have to take on more stringent measures to meet the climate challenge. However, in the short run, when early action is at a premium, a bottom-up approach to climate mitigation may well deliver more and earlier mitigation than top-down approaches.

There are three possible objections to this conclusion that should be addressed head on. First, a bottom-up approach leaves little scope to assess whether the sum total of measures is collectively consistent with meeting the climate challenge. However, if measurability and predictability results in less effective action, there is surely a case for rethinking the approach.

Second, rather than sacrificing predictability, perhaps developing countries should be urged to take on absolute reduction caps. However, an effective climate deal cannot come at the expense of a globally legitimate agreement. Moreover, there is little doubt that a climate regime that locks in dramatically unequal per capita emissions across countries, which a set of absolute caps based on current emission levels would do, would be rejected as unfair by much of the developing world.

Third, a bottom-up approach by itself may fail to satisfy political demands by the North that the South make meaningful commitments to limit emissions. However, this argument conflates effectiveness and the use of targets. If, indeed, a bottom-up approach promises larger and earlier actions, then the onus must be on advocates in the North to reshape the nature of political demands in the North, rather than bend the regime in a direction of lower effectiveness to suit political conditions in the North.

In sum, environmental credibility and predictability in developing country actions are undoubtedly to be desired. However, if the quest for predictability comes at the cost of misaligned incentives and a regime that cannot be made consistent with development realities, there may be good reason to open the door to other approaches. Since it avoids both these problems, a bottom-up approach offers, at least in the short run, an alternative and potentially more effective avenue to early mitigation action by the developing world.

FURTHER READING

J. Gupta, K. van der Leeuw, and H. de Moel, "Climate Change: A 'Glocal' Problem Requiring 'Glocal' Action," *Environmental Sciences* 4, 139–148 (January 2007).

K. Neuhoff, *International Support for Domestic Climate Policies in Developing Countries* (University of Cambridge, 2008).

H. Winkler, R. Spalding-Fletcher, S. Mwakasonda, and O. Davidson, "Sustainable Development Policies and Measures: Starting from Development to Tackle Climate Change," in R. Bradley and K. Baumert (eds.), *Growing in the Greenhouse: Protecting the Climate by Putting Development First* (World Resources Institute, 2002).

Chapter 19

Operationalizing a Bottom-Up Regime
Registering and Crediting NAMAs

Rae Kwon Chung
Ambassador for Climate Change, Republic of Korea

Key Points

- Nationally appropriate mitigation actions (NAMAs) are one type of mechanism to accelerate developing country participation in GHG emissions reduction efforts.
- NAMAs could be purely voluntary (for inherently financially viable projects), internationally supported (for risky or expensive projects), or capable of producing tradable credits (for projects in between these two categories), the latter categories requiring international consensus.
- NAMAs would run their MRV through a central global registry, with greater levels of regulation for projects that are not purely voluntary, especially for those projects that require significant levels of finance up front.
- Tradable credits would draw the lowest cost emissions reductions and bring in large financial flows, assuming that sufficient demand can be realized from market participants and governments in developed countries.

Although the historical burden of climate change rests with Annex I countries, non–Annex I countries are assisting—in their own ways—with mitigating climate change. However, the current structure of the United Nations Framework Convention on Climate Change (UNFCCC) and Kyoto

Protocol allows these contributions to be neither recognized nor coordinated. South Korea has proposed that Nationally Appropriate Mitigation Actions (NAMAs) undertaken by governments be registered with an international NAMA registry, and for appropriate countries and NAMAs, carbon credit or development assistance might be given.

This paper outlines the South Korean NAMA proposal, and in doing so, it aims to address three major concerns related to such bottom-up approaches, as they are likely to be a component of any future climate agreement. The first concern is that a bottom-up approach will be unable to guarantee sufficient reductions to prevent catastrophic warming. The second concern is that a bottom-up approach will not provide sufficient accountability to ensure that carbon finance funds are used effectively. The third concern is that a bottom-up approach will not be able to guarantee that mitigation measures are undertaken as efficiently as possible. The NAMA proposal contains means of effectively addressing all three concerns.

Nationally Appropriate Mitigation Actions

A NAMA could be any action ranging from economy-wide mitigation targets to a specific project in a specific sector. Examples include sustainable development policies and measures (SD-PAMs), reducing emissions from deforestation and forest degradation (REDD), cap-and-trade schemes, sector-wide technology standards, sectoral targets, a carbon tax, building insulation codes, or congestion targets. Nations would be free to choose to undertake as many or as few NAMAs as they would like.

However, NAMAs would only be eligible for carbon credits or other financial support if they fulfilled certain conditions. Consequently, there would be three types of NAMA. The first would be voluntary NAMAs, or those that require no support and do not qualify for credits. The second would be NAMAs that qualify for international support. The standards for determining whether or not a NAMA would receive financial support could be determined by either bilateral or multilateral agreement. The third would be NAMAs which are eligible for carbon credits. The standards for determining whether or not a NAMA would be eligible for carbon credits would need to be consistent with the standards adopted by other carbon markets to allow for linkage to those markets. In theory, a project may fall into both the second and third categories, i.e., it may both qualify for support at the outset and be eligible for carbon credits upon completion.

The boundaries between these three types of categories will need to be determined by some form of international consensus or cooperation. In the absence of such cooperation, all NAMAs are voluntary. As a result, this approach provides substantial flexibility to alter the crediting and support standards to achieve policy goals. The most efficient and effective outcome would be achieved by (a) making NAMAs that will pay for themselves voluntary, (b) providing credits but no additional support for those NAMAs that are not cost-effective on their own but would be cost-effective if credited, and (c) providing additional financial support for valuable mitigation actions that are too risky or expensive to pay for themselves through crediting alone. Under this approach, NAMAs that qualify for carbon credits would likely be the primary mechanism for financing mitigation measures in developing countries.

The crediting of NAMAs could be structured similarly to the current structuring of the Clean Development Mechanism (CDM). Credits would be awarded for reductions below efficiency standards or intensity targets. This adds additional flexibility to the NAMA approach, by allowing parties to set different emissions standards for different projects or sectors. Credits would be purchased by Annex I governments and market participants to meet their targets. The result would be a transfer of financial resources from those countries to mitigation actions in developing countries.

For example, the efficiency standard could be set higher or lower based on different priorities. Alternatively, credits could be issued on the program, policy, or sectoral level. For example, it might make sense to award credits on the project level for LDCs with limited net emissions, but only on the sectoral level for major emerging economies where non-sectoral measures are less likely to be effective.

Although the flexibility provided by this approach will yield many benefits, it will also present some challenges. It may take some time to establish politically acceptable intensity targets or methodologies for all the various NAMAs that are likely to be undertaken. Fortunately, many targets and methodologies could be based off of preexisting CDM practices. Also, not all of the details need to be spelled out immediately. Parties may agree on the principle of crediting NAMAs at Copenhagen, while leaving determinations of appropriate standards for later.

MRV and a Global Registry

For NAMAs to qualify for support or credits, they would have to be done in a measurable, reportable, and verifiable (MRV) manner, as defined by Bali Action Plan paragraph 1(b)(ii). MRV standards will act as the link between mitigation efforts in developing countries and financing from developed countries. As a result, NAMAs should be registered in a central registry charged with keeping track of both the total mitigation actions taken by developing countries and the total financing provided by Annex I nations. The registry would function to ensure that any mitigation measures that receive financing or credits do, in fact, result in verifiable emissions reductions, and that all eligible and reported mitigation measures do, in fact, receive financing or credits.

Registration requirements would be different for each of the three different types of NAMAs. Voluntary NAMAs would not need to be registered, but developing countries should have the option of registering them in an MRV manner in order to keep track of their overall mitigation actions. This information would be valuable in providing an accurate account of developing countries' full contribution to global emissions mitigation, as well as helping determine to what extent developing nations are voluntarily taking cost-effective mitigation actions that do not qualify for credits or support.

NAMAs that require support would need to be registered based on an MRV methodology agreed upon by the party or parties providing the support. This could vary substantially on a project-by-project basis, depending on the priorities of the parties involved.

NAMAs that qualify for credits would be subject to a more stringent MRV methodology. This methodology would need to ensure that credits are effectively equivalent to other carbon credits. This is necessary to allow linkage to global carbon markets and to maintain environmental integrity.

Countries would not receive carbon credits until the project was actually completed, but other forms of support could potentially be issued before a project is initiated or completed. However, firms should still be able to receive financing for the project prior to crediting. The participant firm would submit a project idea to a bank to get loans to initiate the project. The firm would later pay back the loans with the revenue generated by the sale of the carbon credits generated by the project. This approach has already been adopted by many firms initiating unilateral CDM projects, which account for half of all CDM projects.

Although monitoring, reporting, and verification requirements will vary depending on the project, several common features will need to be registered for all projects. Quantity of emissions, support needed and issued, and credits issued should all be registered. Timeframes for project completion could also be registered. Many countries may not have capacity to measure and register their mitigation efforts. Consequently, developing nations should have the option of requesting and receiving assistance to establish the requisite capacity.

The Advantages of Tradable Credits

One of the most important features of this program is the provision of tradable credits. Credits issued for NAMAs would be tradable with firms and nations abroad, similar to CDM, or Emissions Trading System (ETS) on a global scale. This system provides several major advantages over other approaches.

The first is that it allows markets to work to achieve the most efficient reductions. As it is less expensive to reduce CO_2 emissions in developing nations than in developed nations, it is more economically efficient to allow reductions to happen in the least expensive locale. A global trading system is the best mechanism to ensure that capital is most efficiently deployed in these circumstances. According to one model, a global trading system including developing countries could reduce global mitigation costs by 70%.

The second is that a tradable credit scheme will allow for significantly greater overall volume of financial flows and technological transfer than an approach that relies on Official Development Assistance (ODA) or institutional financing. The vast majority of the technology and finance necessary to reduce emissions belongs to the private sector. Consequently, the private sector, not the government, should be the primary source of financial and technological transfer. Furthermore, it will be more politically feasible in Annex I nations to arrange for financial and technology transfers through a trading scheme than through ODA.

For a credit trading scheme to be successful, Annex I will need to ensure that there is sufficient demand for the credits. If the price of credits is too low or unpredictable, then private investors will not have sufficient incentive to invest sufficient money to ensure a net reduction in emissions. There are two primary ways that Annex I nations can make sure that there is sufficient demand for carbon credits. The first is to adopt more ambitious national emissions targets. By adopting more ambitious targets, Annex I countries will force their market participants to purchase more carbon credits in order to meet their targets, which will in turn increase the demand for credits from developing countries. An alternative approach would be for Annex I governments to purchase the credits directly for retirement.

Although both of these approaches would entail substantial costs, they are preferable to the alternative, which would be to finance projects directly through ODA. These techniques would be both more efficient and more effective than increasing ODA, for the reasons described above. They would also be more politically insulated than decisions about direct assistance. Finally, the MRV registry would enable efficient tracking of total financial transfers from Annex I to developing nations.

Conclusion

Any climate change proposal will fail unless it both receives substantial participation from developing countries and creates strong incentives for private companies to invest in mitigation actions. This proposal will create strong incentives for both groups to vigorously participate in global GHG mitigation efforts.

This proposal also contains the tools necessary to achieve the scale, efficiency, and accountability necessary to address global warming. So long as the price of carbon credits is sufficiently high, developing countries should undertake cost-effective mitigation efforts to receive those credits. If efficiency standards are set wisely, then investments should flow to where they will be most effective. And if a central registry is established, then every significant mitigation action and dollar spent on carbon finance will be accounted for.

FURTHER READING

Center for Clean Air Policy, *Sectoral Approaches: A Pathway to Nationally Appropriate Mitigation Actions* (December 2008).

Hyo-eun Jenny Kim, "NAMA Registry—Korea's Proposal for Post-2012 Negotiation," *Presentation at Key Issues of the Post-2012 Climate Change Framework OECD Global Forum on Sustainable Development* (March 2009).

Republic of Korea, *Market-Based Post-2012 Climate Regime: Carbon Credit for NAMAs* (August 2008), available at http://unfccc.int/files/meetings/ad_hoc_working_groups/lca/application/pdf/market_based_climate_regime-korea.pdf.

B

Conditionality and Its Governance

Chapter 20

From Coercive Conditionality to Agreed Conditions

The Only Future for Future Climate Finance

Jacob Werksman

Director, Institutions and Governance Program,
World Resources Institute

Key Points

- A prerequisite to a successful global deal on climate change is the closing of the gap between expectations held by developed and developing nations with regard to the quantity and type of climate finance.
- The significant increase in economic and political power in the developing world is leading to the growing influence of recipient countries on the terms of climate finance.
- The traditional model of conditionalities, whether set at the international or national (investor/developed or recipient/developing) level, needs to yield to a new model in which donors and recipients agree on the conditions under which investments are most likely to succeed.
- Under this new model, a growing recognition of the power of developing countries to set policies and priorities for development finance will be accompanied by greater levels of responsibility and accountability for the way in which investments are made.

A global deal on climate change will depend upon closing the gap in expectations between developed and developing countries on climate finance. Most multilateral environmental agreements (MEAs) provide for the transfer of financial and technical resources from richer countries to poorer countries. These transfers serve the practical purpose of financing developing country capacity to implement projects and policy, and the political purpose of providing incentives for developing country participation in responses to global environmental challenges.

However, financial transfers rarely come without strings attached, i.e., conditionalities imposed by contributor or lending institutions on recipient countries. Conditionalities are thought to be particularly important in the context of global environmental agreements, where scarce financial resources must promote global public goods—such as protecting biodiversity, the ozone layer, and the climate system—that may not be policy priorities for the recipient country. While conditioning access to funds is designed to ensure that the money buys results, it can lead to resentment and a lack of ownership by recipient countries.

In the context of climate change, conditionalities operate in a particularly complex political environment. Climate finance represents, in the eyes of many developing countries and observers, a form of compensation for the damage done to the climate by more than a century of developed country historical emissions, and several decades of continued emissions growth in the context of a growing scientific certainty about the extent of this damage. At the same time, however, the science tells us that even if developed country emissions drop to zero, the growing emissions in the developing world, particularly from emerging economies, will still lead to dangerous climate change.

The climate change negotiations thus raise unique challenges for development assistance. While the South can with some legitimacy demand financial support for reducing emissions, the North and the international community as a whole can legitimately demand a return on this investment. In this context, who gets to set the conditionalities that will, in turn, drive the investments in countries dependent on climate finance?

Broadly, there are three main sources of conditionalities that will determine how climate finance is invested in developing countries:

- Policies agreed multilaterally by the Conference of the Parties to the United Nations Framework Convention on Climate Change (UNFCCC) and any international financial institutions that may be mandated to implement the climate deal

- Policies set unilaterally by national legislation and policies in contributor countries, including mandates on how bilateral assistance will be spent and what kinds of activities will be supported by carbon markets, and/or
- Policies established by the developing country government itself, in the context of national low-emission development strategies and national adaptation plans

Each of these sets of policies emerges from and will shape a dynamic of power, responsibility, and accountability in the relationship between investor contributor governments, investor institutions, and host governments. Ideally, these policies would align, resulting in conditionalities that drive investments that are consistent with nationally determined priorities. In reality, this seems unlikely. Multilaterally agreed policies tend to drift towards a lowest common denominator, as negotiators are required to accommodate the competing concerns of multiple contributors and diverse recipients. Bilateral policies tend to reflect the priorities and interests of the contributor governments, through an exercise of power that favors particular countries, technologies, and policies.

National policies to mitigate emissions and adapt to the impacts of climate change, where they exist, are in the early stages of formation. In developing countries, many of these plans are vague and targeted at an international audience, rather than well grounded in a national consensus. If the latest round of negotiations on climate finance is to succeed in leveraging significant transformations in developing countries, multilateral and bilateral policies will need to support and align with national planning processes. This will require a shift in power from contributor to recipient countries, and a greater sense of responsibility and accountability by recipient countries.

A Brief History of Climate Finance

I will quickly review how the two previous efforts to do a deal on climate finance have distributed power, responsibility, and accountability between contributors and recipients, and then speculate how a new—and better—kind of bargain may be emerging from the Copenhagen process. This summarizes a much longer and more detailed piece of research that the

World Resources Institute is undertaking to both study and inform the negotiations.

The post-2012 climate regime will depend on building upon and agreeing to different terms for climate finance than have been set by the two previous (and largely unfulfilled) climate bargains struck. The first is set out in the UNFCCC, opened for signature at the UN Conference on Environment and Development (UNCED) in Rio in 1992. The Rio Bargain provides, in essence, that the agreed full incremental costs of developing country actions will be financed on a grant basis by developed countries. The bulk of these grants will be transferred through a single financial mechanism, operated by the Global Environment Facility (GEF) under the guidance of the Conference of the Parties to the Convention (COP).

The COP sets the most general of guidance, while operational policies and programs are agreed internationally by the GEF Council. Specific projects are implemented in accordance with the policies of one or more of the GEF's implementing agencies (i.e., the World Bank, the United Nations Development Programme (UNDP), and the United Nations Environment Programme (UNEP)).

Access to GEF funding requires a demonstration that GEF investments represent no more than the "incremental costs" of implementing the Convention and, in doing so, generating a "global environmental benefit" in the form of emissions reductions. In other words, what is funded is by definition that which is not in the national interest. World Bank and UN program officers manage the project cycles and recover their costs through administrative fees. Their environmental and social safeguard standards guide project design and implementation; their financial systems provide for fiduciary accountability. Developing country access to GEF funds is thus mediated conceptually through incremental cost financing, and institutionally through the operations of the intermediary institutions that contributors entrust with designing and overseeing project implementation. This represents a classic contributor-recipient relationship, with conditionalities set and enforced through the exercise of the contributor prerogative. It is not surprising, under these conditions, that GEF projects are often criticized as having little of the catalytic effect necessary to transform national policies and priorities.

The next stage in the development of the climate regime emerged from the negotiations of the 1997 Kyoto Protocol to the UNFCCC, which was designed to put in place the first internationally agreed upon, legally binding cap on greenhouse gas emissions. Developed countries were required to limit their emissions on average to 5% below 1990 levels between 2008 and 2012. With regard to climate finance, the Kyoto Bargain imported the incremental cost concept, as well as the GEF and its supporting institutions.

But in partial response to the observed shortcomings of the financial flows generated by the GEF, the Kyoto Protocol (KP) parties turned to market mechanisms as an additional source of money, incentives, and conditions. The KP's Clean Development Mechanism (CDM) provides a means for incentivizing investments in emissions reducing projects in developing countries by rewarding investors with carbon offsets for each ton of carbon equivalent of emissions reduced (the CDM also provides a source of grant revenue for adaptation activities in developing countries, by providing that a 2% share of the proceeds from each CDM investment be set aside for this purpose). The CDM's Executive Board sets, at the multilateral level, the conditions under which projects are eligible as CDM investments. The host government must agree to the project, but the investor makes the choice of project and investment. Thus, developing countries could exercise the sovereign power to block a project, but in essence, a new specialized global administrative body, and the private sector, determine whether the project is viable.

Fundamentally, the CDM seeks to commoditize the emissions reduced as measured against a business-as-usual baseline—what the GEF would characterize as the "global environmental benefit"—into a tradable return on the investment. Power shifted from the exercise of the contributor prerogative to the operations of a global administrative body overseeing a private-sector market. Developing countries have been frustrated by the slowness with which the CDM's Executive Board has performed its oversight function, and some of the smaller developing countries have been frustrated by the private sector's pursuit of investment in larger industrialized countries where low-cost offset opportunities are easier to come by. Responsibility and accountability for the performance of CDM projects is largely outsourced to the project sponsors and to private-sector companies under contract to monitor and certify the emissions reductions as they occur. It is hard to find evidence that the CDM is promoting investments or incentivizing policy changes at the mainstream in the countries where it operates.

Towards a Successful Climate Change Deal

A successful climate change agreement will depend heavily on a new kind of bargain on climate finance that catalyzes the kind of transformational change in developing countries that previous deals have failed to deliver. There have been promising signs that the dynamic of power, responsibility, and accountability is shifting.

Developing countries, particularly emerging economies, are more economically and politically powerful than during earlier periods of climate negotiations. The size of their economies and the size of their emissions demand greater recognition. While this may not yield significant new financial flows, particularly in the context of a global economic downturn, it does mean that developing countries are likely to demand and receive more formal power in the operation of any new financial mechanisms.

The CDM's adaptation levy, which is collected without reference to contributor purse strings, has led to the creation of an Adaptation Fund Board (AFB) governed by a majority of developing country representatives. It is possible that negotiators will agree to tap new sources of climate finance de-linked from developed country coffers, for example, levies on international air and maritime bunker fuels, and the international auctioning of emissions allowances. This may lead to a mutually agreed relaxation of the contributor prerogative to set conditionalities.

In the context of the Adaptation Fund, and in discussions around any new financial mechanisms established post-2012, developing countries are also demanding greater responsibility for themselves in the programming of climate finance. Submissions call for "direct access" that would allow recipient countries to bypass the traditional "implementing agencies" by nominating national institutions to receive, program, and account for projects funded under a new climate regime. A number of developing countries have expressed their willingness to demonstrate that national finance ministries or planning ministries can meet international fiduciary standards in order to justify this more directly responsible role.

A re-opening of the incremental cost concept may also be on the table. Inspired by efforts that have been made to calculate the marginal abatement costs of reducing emissions in developing countries, some developed countries have suggested that those projects and policies that can be shown to produce near-term positive rates of return should be imputed to a developing country's business-as-usual baseline. In other words, the next generation of climate finance would assume developing countries will discover and invest their own resources in those activities that have both domestic and global environmental benefits. Grants and concessional loans would only be available for investment further up the marginal abatement cost curve. Thus far, developing countries have rejected this approach.

The Copenhagen round of negotiations has also produced some movement in the direction of more direct accountability of developing countries for the investments they host, in exchange for greater accountability of contributors for following through on their financial commitments. The negotiators sketched out the essence of this reciprocal relation in 2007. The Bali Action Plan provided—within the same circuitous sentence—that both "nationally appropriate mitigation actions by developing country Parties" and the "technology, financing and capacity-building" to support and enable these actions must be included in a Copenhagen deal in "a measurable, reportable and verifiable manner."

From Coercive Conditionality to Agreed Conditions

As this book highlights, the costs of a serious response to climate change will likely dwarf the level of development finance available. Moreover, stabilizing the greenhouse gas concentrations in the atmosphere at safe levels will not happen if developing country emissions continue to rise. The dynamics of the negotiations around climate finance have slowly come to recognize this, by searching for new sources of funds, and by beginning to explore ways in which power, responsibility, and accountability for the delivery of climate finance are shared between developed and developing countries.

Coercive conditionalities are profoundly disempowering for developing countries, as they are placed in the position of recipient required to perform against an imposed set of standards. A new and better relationship turns on a recognition that success will depend not on coercive conditionalities, but rather on wise investments that create the right institutional and policy conditions in recipient countries for more sustainable climate-related polices to take root. Direct access to funding for developing countries whose national institutions can demonstrate they meet fiduciary standards, and national systems for measuring, reporting, and verifying funded actions are two new dimensions of a more reciprocal relationship between contributors and recipients that reflect an agreement on the conditions necessary to empower developing countries to shape their own climate policies.

In other words, the next generation of climate finance needs to focus on the incentives necessary to promote good governance within recipient countries—by strengthening the institutions necessary to perform the functions of responsibility and accountability previously performed by intermediary institutions. There are good signs from early efforts to fund activities to reduce emissions from deforestation and degradation (REDD) that both contributors and recipients are recognizing this essential link between governance and effective climate finance.

Both the World Bank and a consortium of UN agencies (UNDP, UNEP, and the Food and Agriculture Organization) are investing in creating the conditions necessary for forest-rich developing countries to combat the drivers of deforestation. The World Bank's Forest Carbon Partnership Facility and the UN-REDD initiative are both providing grants to help these countries demonstrate their readiness to host large-scale forest offset projects by funding assessments of their institutional capacity. These studies are beginning to reveal gaps in countries' capacity and commitment to make and enforce basic land tenure and land use policies, to recognize and uphold the rights and interests of local forest-dependent people, and to police and discourage international trafficking in illegal forest products. Civil society groups in these countries and internationally are taking note, and they are beginning to see REDD, and this new approach to climate finance, as a means of getting citizens involved directly in assessing their governments' readiness to participate in these new deals.

The involvement of national civil society in the design of climate policy is the only means of ensuring that the relatively weak incentives made available by international climate finance can take hold in a way that transforms economies. This requires a shift in power from contributor to recipient government, and then to ultimate beneficiaries of these flows: the communities and the citizens that will host these investments.

FURTHER READING

GEF Evaluation Office, *Fourth Overall Performance Study of the GEF (OPS4) Interim Report*, GEF/ME/C.35/Inf. 1/Rev.1 (2009), available at www.gefweb.org.

D. Reed, *The Institutional Architecture for Financing a Global Climate Deal: An Options Paper*, Technical Working Group, WWF-International (2009).

Chapter 21

Getting Climate-Related Conditionality Right

Kevin E. Davis

Beller Family Professor of Business Law, NYU School of Law

Sarah Dadush

Research Fellow, Institute for International Law and Justice, NYU School of Law

Key Points

- Climate-related conditionality is an inevitable feature of many future public and private investments in developing countries.
- Climate conditionality raises two basic sets of substantive concerns, one relating to its effectiveness and efficiency, the other to conflicts between climate conditionality and broader development goals and equitable concerns.
- In order to ensure effectiveness and consistency with other objectives, publicly funded investment funds using climate conditions should provide due process to the citizens of recipient countries.

Conditionality has gotten a bad name in development finance. But it may be rehabilitated by the emerging climate change regime. Mitigating climate change by reducing emissions of greenhouse gases (GHGs) from developing countries will require substantial amounts of capital. Some of that capital will come from individuals or organizations who insist that their funds be used in ways that tend to promote mitigation. In other words, they will insist on conditionality. This raises a number of policy concerns, including several that are reminiscent of debates about conditionality in other contexts.

The first part of this paper provides an overview of existing forms of climate-related conditionality. The second part sets out the main substantive issues involved. The third part considers implications for institutional design and the process by which conditions are formulated.

The Landscape of Climate-Related Conditionality

Climate-related conditionality can take a number of different forms, ranging from obligations for the recipient of funds to reduce emissions from its own activities, to obligations to encourage other actors to reduce emissions, to obligations for recipients to report on their own or others' efforts to mitigate climate change. Many different kinds of organizations have demonstrated interest in imposing conditionality of one sort or another on financial transfers to developing countries or to enterprises or projects located in those countries.

Public Funds Dedicated to Mitigation

A number of large funds sponsored by public actors have been created to channel mitigation-related capital to actors in less-developed countries on concessional terms. These funds are dedicated exclusively to investments in mitigation. Funds created under the auspices of the United Nations Framework Convention on Climate Change (UNFCCC) and the Kyoto Protocol and through other multilateral initiatives include:

- Global Environmental Facility (USD 3.1 billion for 2006–2010)
- United Nations Collaborative Programme on Reducing Emissions from Deforestation and Forest Degradation in Developing Countries (UN-REDD) (USD 35 million)
- World Bank—Forest Carbon Partnership Facility (USD 165 million)
- World Bank—Climate Investment Funds (USD 6.1 billion), made up of the Clean Technology Fund and the Strategic Climate Fund

Additionally, instead of financing specific projects, the World Bank Carbon Finance Unit (CFU) uses money contributed by governments and companies in Organisation for Economic Co-operation and Development (OECD) countries to purchase project-based GHG emission reductions in developing countries and countries with economies in transition. The reductions are purchased through one of the CFU's carbon funds on behalf of the contributor and within the framework of the Kyoto Protocol's Clean Development Mechanism (CDM) or Joint Implementation (JI) program.

Bilateral initiatives by developed country governments include:

Japan—Cool Earth Partnership (USD 10 billion)
UK—Environmental Transformation Fund (GBP 800 million)
Norway—Climate and Forest Initiative (€ < 600 million)
United Nations Development Programme—Spain MDG Achievement Fund (€90 million)
EC—Global Climate Change Alliance (€100 million)
Germany—International Climate Initiative (€400 million)
Australia—International Forest Carbon Initiative (AUD 200 million)

Other Bodies That Have Adopted Climate-Friendly Standards and Investment Policies

While many organizations that invest in developing countries do not have funds dedicated exclusively to investments in mitigation, they have adopted policies that call for giving priority to climate-friendly investments or at least for avoiding investments that have the opposite effect. Some of these policies are legally binding, others are voluntary.

Publicly sponsored organizations that have taken steps to incorporate climate change concerns into their investment decisions include the International Finance Corporation (IFC), the Multilateral Investment Guarantee Agency (MIGA), and the World Bank, all of which include the reduction of GHG emissions among the priorities they seek to advance in their financing of projects.[1] As a result, these organizations often make financing of projects conditional on the climate-friendliness of those projects.

Several associations of financial intermediaries have adopted voluntary codes of conduct that include commitments to support only climate-friendly projects. One such initiative is the Equator Principles, which have been adopted voluntarily by over 60 project finance institutions. The Principles require participating institutions to observe the IFC Performance

Standards in their lending activities and to provide annual reports on their progress. The IFC's Performance Standards currently require, among other things, clients to report certain GHG emissions and encourage them to employ cost-effective measures to reduce or offset emissions.[2]

Another example is the Investor Network on Climate Risk (INCR), a network of more than 80 leading institutional investors with collective assets of more than USD 7 trillion. In 2008, INCR announced its Action Plan calling for investors to take nine specific steps to address the growing risks and opportunities from climate change, with a significant focus on reducing GHG emissions. The steps include the following commitments:

- Support clean technology, with a goal of deploying USD 10 billion collectively over the next two years
- Require and validate that investment managers, investment consultants, and advisors report on how they are assessing climate risks in their portfolios, including risks from new carbon-reducing regulations, physical impacts, and competitive risks
- Encourage Wall Street analysts, rating agencies, and investment banks to analyze and report on the potential impacts of foreseeable long-term carbon costs in the range of USD 20 to USD 40 per metric ton of CO_2, particularly on carbon-intensive investments such as new coal-fired power plants, oil shale, tar sands, and coal-to-liquid projects
- Push the SEC to issue guidance leading to full corporate disclosure of climate risks and opportunities

Substantive Considerations

The consequences of adopting any given form of conditionality can be evaluated along a number of dimensions. First, will the conditions be effective? In other words, are the expectations that climate-related conditionality will have a significant effect on the behavior of potential recipients of funding—or other actors—justified? Second, will the resulting reductions of GHG emissions be cost-effective? Third, what impact will climate-related conditionality have on economic development in developing countries? Fourth, will this form of conditionality promote or undermine the equitable distribution of wealth and economic opportunity, either across or within countries? We consider each of these questions in turn.

Effectiveness

The idea that climate-related conditionality will exert a meaningful influence on behavior cannot be presumed. No particular form of conditionality will be effective unless it is adopted by enough investors to cause a meaningful reduction in the amount of capital available without the relevant conditions. A single bank's refusal to finance coal-fired power plants will have little or no effect on overall investment in that type of project.

Accordingly, the most successful conditions, in terms of effectiveness, are likely to be ones attached to funding provided by the multilateral and regional development banks. Those entities remain an important source of funding for many developing countries, especially when it comes to concessional funding targeted at mitigation and adaptation-related projects. Moreover, through the Equator Principles and similar initiatives, the conditions imposed by the development banks also tend to be adopted by large numbers of private actors—a form of cross-conditionality.

Of course, the effectiveness of any given set of conditions depends on whether they are actually enforced. It is not always in the interests of funding organizations to insist upon compliance with climate-related conditions. For instance, managers of profit-oriented funds that have signed on to the Equator Principles may still be tempted to invest in carbon-intensive projects that offer high economic returns. Meanwhile, employees of development banks may experience pressure to fund dirty projects, either from member states or from constituencies within their organization who have an interest in maximizing the volume of lending.

Effectiveness is a question that would benefit from empirical research. It would be useful to know, for example, whether organizations that have the right to insist on compliance with climate-related conditions either ignore instances of non-compliance or waive the right to insist on compliance. If organizations do relax their compliance or enforcement standards, it would be helpful to know when and why they do so.

Cost-Effective Emission Reduction

To the extent that climate-related conditionality is an effective method of altering the behavior of the recipients of funding or other actors, the next question is whether the result is a cost-effective reduction of GHG emissions. There are several reasons why this outcome cannot be presumed. First, some funds may employ social or economic conditions—

including sectoral or geographic limitations—that go beyond requiring emission reductions and preclude investment in projects associated with relatively cost-effective reductions in GHG emissions. Second, when a number of projects satisfy the conditions of a given fund, it may, either advertently or inadvertently, fail to give priority to the projects that offer the greatest reduction in GHG emissions per unit of capital invested.

One way to address these concerns is for funding organizations to review their conditions regularly to ensure that they are promoting cost-effective emission reductions. A more fundamental response would be to abandon funding conditionality altogether and rely on economic actors to identify cost-effective mitigation opportunities using the price signals generated by, say, a cap-and-trade or credit trading system.

An additional consideration is that conditionality entails certain transaction costs—the costs that both providers and recipients of capital bear in monitoring, reporting, and verifying compliance with any given set of conditions. Those costs can be particularly significant for developing countries with limited institutional capacity. Transaction costs may also be particularly high when recipients have to comply with several distinct sets of climate-related conditions. If the benefits of conditionality were outweighed by the related transaction costs, this would weigh in favor of abandoning conditionality (although the transaction costs associated with alternatives may not be trivial). One strategy for limiting the costs of conditionality is to enhance consistency across the climate-related conditions imposed by various financial institutions, both public and private. This could be accomplished through explicit harmonization, incorporation by reference to international standards developed through the Copenhagen process, or forms of cross-conditionality where one organization adopts another's standards.

Host-Country Development

Allocating capital in a fashion that efficiently reduces GHG emissions is not necessarily consistent with maximizing benefits to society along other dimensions. In the absence of regulation, the most climate-friendly projects are usually not the ones that generate the largest pecuniary returns for investors. Likewise, climate-friendly projects will not necessarily generate the greatest amounts of employment, the most helpful forms of technology transfer, or the most effective forms of adaptation to climate change. This raises the potential for conflicts between the interests of actors concerned primarily with climate change mitigation and the interests of inhabitants of developing countries.

This issue is coming to a head in the debate over whether the multilateral development banks and other financial institutions should finance coal-fired power plants. The World Bank has a goal of having 50% of its energy portfolio dedicated to low-carbon investment (which includes clean coal with several conditions attached). If enforced, this policy will reduce the supply of capital for new coal-fired power plants to some extent. Is this in the best interests of countries that desperately need cheap energy to sustain their economic development? These concerns are particularly pressing for the Least Developed Countries, which desperately need growth and are only minimally responsible for past and present GHG emissions, and yet are also most vulnerable to the negative consequences of global warming.

Equity

The benefits of GHG emissions reductions will be distributed globally, though not necessarily uniformly. Meanwhile, to the extent that individual projects create jobs, transfer technology, or support adaptation to climate change, the costs and benefits are likely to be concentrated in the projects' host countries, and even among particular segments of society. Conditions that preclude financing for coal-fired power plants are one example: they may provide global benefits at the expense of the inhabitants of developing countries. As another example, conditions that promote investments in REDD may produce globally diffused benefits in the form of climate change mitigation and locally concentrated benefits for governments or private landowners who receive cash transfers to encourage forest conservation. But these conditions may simultaneously impose substantial costs on indigenous groups prevented from using forests in traditional ways.

Another important issue is whether it is appropriate to impose conditions that are inconsistent with the distribution of mitigation-related costs agreed to by states in international negotiations. In other words, is Copenhagen-plus conditionality acceptable? This question is likely to be a particularly pressing one for the multilateral development banks, whose conditionality arguably should not deviate significantly from international law.

Implications for the Process of Implementing Conditionality

The processes by which conditions are formulated and enforced also raise some extremely important concerns. Due process in conditionality—in the colloquial rather than the legal sense—is intrinsically worthwhile, and may also, to the extent that it enhances legitimacy, tend to induce both providers and recipients of capital to adopt and comply with conditions. As we have already argued, widespread adoption and compliance is important if conditionality is to be effective and implemented with minimal transaction costs.

In the context of climate-related conditionality, the central procedural questions revolve around the roles that different parties, especially the inhabitants of recipient countries, ought to play in formulating conditions. These questions are particularly important for conditionality imposed by publicly sponsored actors. It seems intuitive that local constituencies affected by the decision of a public actor ought to be entitled to benefit from well-designed accountability, transparency, and participation mechanisms. In other words, to the extent that a fund's investment decisions affect the level or distribution of wealth in a society, it ought to be accountable to members of that society who in turn ought to be able to participate in those decisions, observe the processes by which they are made, and hold the decision-makers accountable.

The difficulty, however, with granting procedural entitlements to actors from recipient countries is that they may favor different substantive outcomes than providers of capital. For instance, they may prefer projects that generate local employment to projects that efficiently reduce emissions. Or they may prefer projects that support adaptation over those that support mitigation. Consequently, granting local actors robust procedural entitlements risks alienating financiers with opposing preferences. Generating this kind of local ownership of the process of formulating and enforcing conditions, without undermining other objectives, is one of the central challenges associated with all forms of conditionality.

Conclusion

Climate-related conditionality in development finance is probably inescapable. The challenge going forward will be to fashion conditions that balance potentially competing interests in effectiveness, cost-effective emissions reductions, development, and equity. Formulating institutions and processes capable of resolving these issues in a legitimate fashion ought to be a central concern in designing a global regime to address climate change.

FURTHER READING

Sarah L. Babb and Bruce G. Carruthers, "Conditionality: Forms, Function and History," *Annual Review of Law and Social Science*, Vol. 4: 13–29 (December 2008).

Charles Di Leva, "Global Warming Litigation Impels U.S. Financing Agencies to Acknowledge Climate Change," in *The Nature of Law: The Newsletter of LEGEN* (March 2009).

The Equator Principles, available at http://www.equator-principles.com/documents/Equator_Principles.pdf.

IFC, "Performance Standards on Social and Environmental Sustainability," available at http://www.ifc.org/ifcext/sustainability.nsf/AttachmentsByTitle/pol_PerformanceStandards2006_full/$FILE/IFC+Performance+Standards.pdf.

Multilateral Investment Guarantee Agency, "Performance Standards on Social and Environmental Sustainability," available at http://www.miga.org/documents/performance_standards_social_and_env_sustainability.pdf.

Gareth Porter, Neil Bird, Nanki Kaur, and Leo Peskett, *New Finance for Climate Change and the Environment* (World Wildlife Federation and Heinrich Boll Stiftung, July 2008).

NOTES

1. See IFC, "Performance Standards on Social and Environmental Sustainability," Standard 3, http://www.ifc.org/ifcext/sustainability.nsf/AttachmentsByTitle/pol_PerformanceStandards2006_full/$FILE/IFC+Performance+Standards.pdf; and World Bank Group, "Environmental, Health, and Safety Guidelines" (known as the "EHS Guidelines"), section 1.1, "Air Emissions and Ambient Air Quality," http://www.ifc.org/ifcext/sustainability.nsf/AttachmentsByTitle/gui_EHSGuidelines2007_GeneralEHS/$FILE/Final+-+General+EHS+Guidelines.pdf.

2. IFC, "Performance Standards on Social and Environmental Sustainability," Standard 3(11).

Chapter 22

Making Climate Financing Work

What Might Climate Change Experts Learn from the Experience of Development Assistance?

Ngaire Woods

Professor of International Political Economy;
Director of the Global Economic Governance Programme,
University of Oxford

Key Points

- Financing for climate change mitigation will likely involve some form of conditionality, although conditionality alone is a poor guarantor of project or policy success.
- The most important factors in determining project or policy success is the alignment of a project with the priorities of the community and local ownership, local implementation, and the timing, certainty, and reoccurrence of funding.

At the heart of any global deal on climate change lies a compact between wealthy and less wealthy countries. The vast majority of industrialized countries have already accepted binding commitments to reduce their greenhouse gas (GHG) emissions (although few have made any progress to reducing emissions in practice). Future progress in limiting emissions relies upon wealthy countries meeting their commitments and—equally importantly—upon major emerging economies agreeing to accept limits on their future emissions. No such deal has yet been forged. At the same time industrialized countries and emerging economies have agreed that support must be offered to poorer developing countries that will be severely affected by the failure (to date) to mitigate emissions. Most proposals envisage that wealthy countries will persuade developing countries by putting financing on the table. But how, and with what strings or performance criteria attached?

Climate change experts have framed the problem as one of how to best use finance to incentivize developing countries to undertake significant near-term mitigation measures and eventually transition to emissions caps. The assumption is that by setting up incentives for developing countries to deliver, the industrialized countries will be able to transform policies and practices in developing countries. Wealthy countries will set goals and disburse financing only upon proven performance of actions taken towards the goals.

A long history of donor efforts to incentivize policymakers in developing countries could usefully inform climate change proposals. Donors often believe that the use of structured incentives and conditionality are sufficient to ensure that countries will adopt particular policies or meet certain objectives. Although this is intuitively appealing, the history of attempts to do this suggests that this is a fundamentally mistaken view. In fact, the main impact of such structured incentives or conditionality may well lie in the effect on the behavior of those providing the financing.

Reliance on incentives and performance-based conditions offers a tempting shortcut, which often leads donors away from examining other elements that are significantly more likely to shape whether or not a government or local authority will achieve particular goals. Let me outline some of these elements.

1. Alignment and Ownership (and How to Test for It)

A core lesson from development assistance is the importance of aligning any foreign-supported proposed policy or project with a country or a community's own priorities and objectives. This lesson has been accepted by major industrialized country donors in the 2005 Paris Declaration on Aid Effectiveness and the subsequent 2008 Accra Agenda for Action. The lesson is also often expressed in terms of ownership: the more a policy or project is owned by those who implement it, and the more closely a project or policy reflects local priorities, the more likely it is to succeed. This goal is easy to state but difficult to translate into operational guidelines. How does one test whether or not a project or policy is locally owned or sufficiently aligned and what does that mean? Important indicators of local ownership could include

- the origination of the project or policy (who had the idea?);
- the design of the project or policy; and
- the financing and resourcing of the project or policy (have locals contributed resources?).

The latter is perhaps the clearest indicator of how much priority a community gives to the idea proposed.

All that said, any tests of ownership require an initial answer to the question "owned by whom?" Is it ownership by a government that matters, or by an individual Minister within the government, or by a disenfranchised minority? Here providers of external financing have to make explicitly political choices about whose visions and aspirations they are supporting within a society. No policy will succeed without local champions—no matter how much this fact might be obscured by the design of performance-based or incentivizing systems. The political choices involved are difficult and complex.

It is even more difficult in practice for donors to support rather than overwhelm local champions. This takes me to a second condition for success.

2. *Local Implementation* (*and Resisting the Temptation to Micro-Manage*)

Implementation must always rely upon local actors and institutions. The experience of aid demonstrates how tempting it is for external funders —who have identified local champions—immediately to use them as a leverage point to try to shape ever wider and deeper areas of policy. For example, some donors who established an initial relationship with communities by helping governments to phase out user-fees for education, have then tried to use these openings to push for other kinds of reform (such as public-private structures) in the education sector. To quote one aid official participating in an Oxford workshop on aid negotiation and management: "it's just really hard for us not to get in there and try to shape everything."

What begins as external support for a local initiative can quickly become a circumvention of local expertise and institutions. This may result in an erosion of local governance, accountability, and the likelihood of success. To prevent such consequences, donors should seek to ensure clarity among themselves and holding one another to account in defining (and limiting) their goals and subsequent influence over implementation.

3. The Timing of Financing

Often the timeline of development disbursal is determined by the donor, which results in aid being delivered either too fast or too slowly. Disbursement pressures to deliver too fast exist where an agency has an annual cycle of lending and its officials need to ensure that they lend out all that is available. By contrast, many agencies deliver aid too slowly where risk-aversion in the bureaucracy and overall risk-minimizing decisionmaking structures result in bureaucratic delays in disbursement. In either case, the timing of financing will greatly limit the likelihood of project or policy or policy-reform success.

4. The Certainty and Recurrence of Funding

Many of the policies and projects aimed at addressing climate change require long-term planning and investments. If governments are to consider external financing in planning for the future and in investing in infrastructure or personnel, the financing must be both certain and recurrent. Yet often aid is both volatile and unpredictable. Performance-based conditions and targets (some of which may not be reached due to exogenous shocks beyond the control of a government—such as a drought, or a global financial crisis) are likely to make projections of aid receipts yet more uncertain. Aid-dependent governments work within very tight constraints. Their spending plans are typically developed within parameters set by the IMF and World Bank. The multilaterals analyze the sources of the government's revenue and its capacity to service debt in the future. These debt sustainability analyses were created to ensure that efforts by governments to meet the Millennium Development Goals do not build up unsustainable levels of future debt. Such a debt build-up would be the result of governments spending money which donors had promised but did not disburse.

5. Reporting Structures and Local Accountability

Providers of external finance usually require direct reporting back to them in forms that fit their own exigencies. For example, each donor government often has its own accounting format that it must use to report to its own auditor-general or parliament. As a result of this, donors require developing country governments to use numerous different formats for reporting. The devastating results have been documented in a study commissioned by donors concerned with this problem. Reporting in this way maximizes the burden on developing countries and does not support efforts within developing country governments to simplify, streamline, and make more transparent their own finances. For these reasons, the reporting structure of external financing is likely to affect the long-term sustainability and accountability of policies or projects.

6. The Adaption and Renewal of Externally Funded Projects or Policies

Finally, projects and policies need constant ongoing adaptation and redesign, as well as formal evaluation and renewal. Part of ownership of a project or policy is ownership of the processes of review, adaptation, and renewal. All too often in development assistance, these processes are conducted by the external funding agency rather than by (or with) the local champions and implementers. The result is a system of reporting and evaluation that is unlikely to bring to light problems which need resolving or redesigning around. Donor-designed adaptation and renewal are also unlikely to strengthen local governance and policymaking capacity.

FURTHER READING

Elinor Ostrom et al., "Aid, Incentives, and Sustainability," *Sida Studies in Evaluation 02/01: Aid, Incentives, and Sustainability: An Institutional Analysis of Development Cooperation* (2002).

World Bank/IMF, *How to Do a Debt Sustainability Analysis in Low Income Countries* (2005), available at http://siteresources.worldbank.org/INTDEBTDEPT/Resources/DSAGUIDEv7.pdf.

Part IV

National Policies

Implications for the Future Global Climate Finance Regime

Chapter 23

Climate Legislation in the United States

Potential Framework and Prospects for International Carbon Finance

Nathaniel O. Keohane[1]

Director of Economic Policy and Analysis, Environmental Defense Fund

Key Points

- There is an increasing prospect of a comprehensive US cap-and-trade program to control GHG emissions. A US carbon market would have crucial implications for carbon finance, particularly in developing countries.
- US climate legislation passed by the House of Representatives in June 2009 would impose a cap-and-trade system to ratchet down GHG emissions from a broad range of sources. Once fully underway in 2016, the cap would cover nearly 85% of US GHG emissions. The cap would reduce emissions from covered sectors by 17% below 2005 levels by the year 2020, and by 83% below 2005 levels (80% below 1990 levels) by the year 2050. These targets, applied to a program of such broad scope, would represent the most ambitious effort to date taken by any country to reduce GHG emissions.
- The proposed legislation provides several points of entry for other countries to gain access to the US carbon market. This would help anchor a new bilateral approach to expanding carbon markets that can be a useful complement to the international agreement being negotiated under the auspices of the United Nations Framework Convention on Climate Change.

- In particular, US GHG allowances would be fully fungible with credits from other emission trading systems with absolute caps on emissions, and similarly stringent monitoring and verification protocols. Moreover, 1 to 1.5 billion tons of emissions from covered US sources could be offset by verified emissions reductions elsewhere in the world, in the form of credits for reduced emissions from tropical deforestation, credits for sector-wide emissions reductions in developing countries, and project-based offset credits such as Certified Emissions Reductions under the Clean Development Mechanism.
- Initially, two-thirds of allowances would be allocated for free, another one-sixth given to state governments to fund energy efficiency, clean tech research, and adaptation. Over the life of the program, 4% of the value of the allowances (an estimated USD 50 billion in present value) would be put aside to fund reductions in tropical deforestation, with another 5% (USD 70 billion) for international adaptation and international clean technology transfer.

On June 26, 2009, the US House of Representatives passed a sweeping bill that would reduce US greenhouse gas (GHG) emissions by 17% below 2005 levels by 2020, and 83% below 2005 levels by 2050. If the momentum from the House bill can be carried on through the Senate, the United States may at last be taking on meaningful domestic action, on the eve of the international negotiations in Copenhagen in December 2009. This chapter sketches the key features of the House bill, focusing on the provisions that would allow linkages between the US carbon market and emissions reduction efforts in other countries.

Overview of the House Bill

The American Clean Energy and Security Act (ACES), sponsored by Henry Waxman (Democrat of California) and Ed Markey (Democrat of Massachusetts), is a comprehensive energy and climate bill. At its heart is a cap-and-trade program that would put a declining limit on allowable GHG emissions from most of the US economy, including all CO_2 emissions from fossil energy use as well as process emissions (of CO_2 and other GHGs) from large industrial facilities. (A separate cap would limit the import and consumption of hydrofluorocarbons.) Sources covered by the cap would be required to submit one allowance for each ton of GHG emissions in each year; the allowances would be fully tradable and bankable. Covered sources could also meet their compliance obligations by purchasing credits for verified emissions reductions from the US forestry and agricultural sectors, as well as from international sources (as discussed later in the chapter).

The cap would take effect in 2012, covering the electric power sector and transportation fuel producers, together accounting for roughly two-thirds of US emissions in 2005. Fuels producers would be responsible for the carbon content of their fuels—that is, the eventual tailpipe emissions from combustion. The cap would be extended to cover major industrial sources in 2014, increasing coverage to just over 75% of 2005 baseline emissions; it would be fully phased in by 2016, when the inclusion of natural gas would increase coverage to nearly 85% of baseline emissions. The cap would decline over time; in 2012, it would limit emissions by covered sources to 97% of their 2005 levels, declining to 83% in 2020, 58% in 2030, and 17% in 2050.

In the initial years, roughly two-thirds of allowances would be allocated gratis to regulated emitters or energy consumers; one-sixth would be allocated to State governments and other non-emitters to fund energy efficiency, clean energy research, adaptation, reductions in tropical deforestation, and other public purposes; the remaining one-sixth would be auctioned by the federal government, with most of the proceeds going to fund tax credits for low-income households. By 2035, the free allocation would be almost entirely phased out. Over the span of the program, the bill would set aside 4% of cumulative allowances (worth an estimated USD 50 billion in present value) to fund reductions in tropical deforestation, along with another 5% of allowances (USD 70 billion) to finance international adaptation and international clean technology transfer.

In addition to the cap-and-trade provisions, ACES contains a range of complementary measures designed to spur energy efficiency and renewable energy, including a combined renewable energy/energy efficiency standard for the electric power sector; strengthened energy efficiency standards for buildings and appliances; performance standards on new coal-fired power plants; and performance standards for industrial sources of emissions below the minimum threshold needed to qualify under the cap-and-trade program. Taken together, the entire bill—including the cap-and-trade provisions, the complementary measures, and supplemental reductions from tropical deforestation achieved through allowances set aside for that purpose—is designed to reduce net US GHG emissions by 20% below 2005 levels by the year 2020, falling to 83% below 2005 levels by 2050.

While the US has lagged behind other developed countries in its commitment to reduce GHG emissions—most noticeably by choosing not to sign the Kyoto Protocol—the proposed legislation represents a sea change. If enacted, it would become the most ambitious GHG emissions reduction program anywhere in the world.

Although the required reductions appear less stringent than those already adopted by the EU, when compared to the 1990 baseline commonly used in international negotiations, that baseline obscures the fundamental changes in the structure of Europe's economies that have occurred since 1990. These include the economic collapse in countries of the former Soviet Union and the ensuing decrease in emissions throughout Eastern Europe, the reunification of Germany (and the subsequent shuttering of highly polluting, inefficient East German factories), and the deregulation of the electric power sector in the United Kingdom (with its attendant dramatic fall in Britain's reliance on coal-fired electricity generation). When compared to the proper counterfactual—emissions in the absence of climate policy—the targets embodied in the proposed US legislation turn out to be at least as stringent as the current EU target.

In addition, the US legislation caps GHG emissions through the middle of the century, specifying a cap for every year through 2050. (In contrast, the current commitment protocol under the Kyoto Protocol lasts only through 2012.) Such a long time horizon is crucially important to provide the clear signals needed to shape investment decisions.

International Linkages under the Proposed Legislation

Of particular relevance, from the perspective of international carbon finance, are the provisions in ACES that would establish links between the US carbon market and emissions reduction or abatement activities in other countries. These various links can be thought of as distinct points of entry into the US carbon market. First, the bill provides for unlimited linkage (full fungibility of allowances) with emission trading systems in other countries, as long as those countries impose mandatory absolute tonnage limits on total GHG emissions at the national or sectoral level, and establish provisions for monitoring, enforcement, and offset quality that are at least as stringent as those under the US program. (Thus allowances issued under the European Union's Emissions Trading Scheme could almost certainly be tendered for compliance with the US system.) The determination of which countries would be eligible for this linkage provision is assigned to the Environmental Protection Agency (EPA), in consultation with the Secretary of State.

Second, ACES would allow covered entities to offset up to 1 billion tons of emissions annually, using credits for verified emissions reductions in developing countries (with the limit applied on a pro-rated basis for individual emitters); as many as 1.5 billion tons could be offset with international credits if the supply of domestic offset credits were limited. Beginning in 2017, international credits are subject to a 20% discount, meaning that emitters must submit 5 credits to offset 4 tons of emissions. These international credits fall into three categories:

1. *International forest credits.* The bill authorizes the EPA Administrator, in consultation with the Secretary of State and the Administrator of the US Agency for International Development (USAID), to enter into agreements or arrangements with countries on reducing emissions from deforestation. To be eligible for crediting, forest nations will have to demonstrate, beginning 5 years from the start of the program (extendable 8 more years in the case of small-emitting and least developed countries), reductions in total emissions from deforestation nationwide, or in their large-emitting states or provinces, from a baseline that results in zero net deforestation within 20 years. Programs in forest nations must be undertaken in compliance with rigorous monitoring and accounting standards, and in consultation with local communities, indigenous peoples, and other stakeholders. In addition, the bill sets aside 5% of the total US allowance pool to assist tropical forest nations in preparing to participate in this program, to preserve existing forest stocks, and to achieve supplemental reductions of 720 million metric tons in 2020, and cumulative reductions of 6 billion metric tons by 2025.
2. *Sectoral credits.* The bill directs the EPA, in consultation with the Secretary of State, to identify sectors and countries that are suitable for crediting on a sectoral basis—meaning that credits are awarded

only if the sector as a whole reduces emissions below a specified baseline. Sectoral crediting would apply to developing countries with high GHG emissions and comparatively high levels of income, for sectors that would be capped if they were in the United States. To gain access to the US market through this sectoral crediting program, listed nations would have to establish a domestically enforceable sectoral baseline of absolute emissions, set at levels below business as usual and consistent with a goal of limiting global warming to 2°C relative to preindustrial levels (equivalently, limiting atmospheric GHG concentrations to 450 ppm CO_2e).

The clear and explicit insistence on meaningful absolute baselines measured in tons—as opposed to no-lose sectoral intensity targets measured in tons per unit of output—is a crucial feature of the legislation. Compared with intensity-based measures, absolute baselines provide more certainty over the resulting emissions, are less susceptible to subsequent manipulation (under intensity targets, allowable emissions depend on measures such as sectoral output, which is itself often imprecisely measured), and prepare developing countries eventually to establish a domestic cap-and-trade system of their own and become full participants in global carbon markets.

3. *Credits issued by an international body.* The bill authorizes the EPA, in consultation with the Secretary of State, to issue offset credits in exchange for international offset credits issued by a body established pursuant to the UN Framework Convention on Climate Change (UNFCCC)—as long as the EPA Administrator determined that the international body's procedures provided equal or greater assurance of the integrity of offsets relative to the US domestic offset program. For example, Certified Emissions Reductions issued under the Clean Development Mechanism (CDM) could be sold into the US market and used for compliance (subject to the 5:4 ratio on international offset credits), at least for project types for which the CDM methodologies for additionality and verification of emissions reductions were found adequate.

Starting in 2016, EPA may not issue project credits for projects in countries and sectors on the sectoral crediting list. This provision is important to create the right incentives for developing countries to move away from project-based offsets—where concerns about additionality and measurement are endemic—and towards sectoral caps.

A Two-Track Approach to Expanding Carbon Markets

One of the most significant aspects of the US legislation is its role as a harbinger of a new model for expanding carbon markets. The bill passed by the House of Representatives, with its multiple points of entry into the US carbon market, represents a parallel track to building international participation in reducing GHG emissions—one that can operate alongside (and as a complement to) the international negotiations under the auspices of the UNFCCC. The UNFCCC provides the multilateral forum for overarching tasks such as setting global targets for emissions reductions and constructing international financing arrangements for clean technology and adaptation. ACES, on the other hand, would establish a bilateral process by which the US could grant access to its carbon market to specific countries—in the form of credits for tropical deforestation, or for sector-wide emissions reductions, or for project-based offsets.

Some critics will surely argue that allowing individual countries to set the rules for access to their carbon markets will undermine efforts to build a single global market. A parallel US system for approving offset credits would likely require a separate set of criteria for additionality as well as for monitoring, reporting, and verification protocols, potentially adding a further hurdle to the development of emissions reductions projects in developing countries. The benefits of a parallel track, however, would outweigh any drawbacks. First, the ability of developed countries to set the terms of access to their markets will be crucial from a political perspective. In the US Congress, for example, one of the most important issues in the debate about climate legislation is whether and when major developing economies like China and India will accept binding limits on their emissions. By setting a high bar for emission reduction credits from other countries, Congress can create the incentive structure that will encourage other countries to act—and therefore help to assuage domestic concerns about taking action in the absence of commitments from other countries.

Second, a parallel approach offers an additional mechanism to encourage (and reward) emissions reductions in major emitting developing countries, in the event that international negotiations are deadlocked or delayed.

Finally, a parallel approach has the potential to create a virtuous cycle in carbon markets. A country that imposes relatively stringent criteria will limit the supply of credits available in its market, effectively driving up their price—creating a stronger incentive for developing countries to meet the more stringent criteria. In effect, this is a carbon-market version of the so-called "California effect"—the positive dynamic that occurs when one (sufficiently large) jurisdiction imposes a higher standard on products and thereby raises the bar for an entire industry.

The US stands on the cusp of a landmark achievement. Climate legislation along the lines of what has already passed the House of Representatives would vault the US to a position of leadership in the international arena, after over a dozen years of lagging behind. As importantly, US action on climate has the potential to induce leading developing economies to reduce their emissions in order to sell credits into a US cap-and-trade system—keeping costs low for American consumers and businesses while securing meaningful emissions reductions around the world and preparing the groundwork for a truly global carbon market.

FURTHER READING

Judson L. Jaffe and Robert N. Stavins, *Linking a U.S. Cap-and-Trade System for Greenhouse Gas Emissions: Opportunities, Implications and Challenges* (Reg-Markets Center, 2008).

Nathaniel O. Keohane, *Oral Testimony before the Subcommittee on Energy and Environment, Committee on Energy and Commerce, United States House of Representatives,* June 9, 2009, available at http://www.edf.org/documents/9924_Keohane%20Testimony%20EC-EE%206-9-09.pdf.

NOTES

1. The author gratefully acknowledges helpful comments from Peter Goldmark, Jennifer Haverkamp, Annie Petsonk, Dick Stewart, and Gernot Wagner on previous drafts of this chapter, while holding them blameless for any remaining errors.

Chapter 24

The EU ETS
Experience to Date and Lessons for the Future

James Chapman
Research Fellow, Center on Environmental and Land Use Law, NYU School of Law

Key Points

- The EU has learned valuable lessons about the effective structure and governance of carbon markets through its experience with Phases I and II of the EU ETS. The scheme has moved the EU towards expected full compliance with its Kyoto commitments. Plans for post-2012 provide for higher levels of auctioning, wider coverage of gases and industry, and a more centralized and harmonized scheme overall.
- There has been some difficulty—especially recently—in creating a sufficiently high price on carbon to incentivize a significant shift towards a low-carbon economy, but a combination of a lower cap for Phase III that decreases annually after 2020 (promising long-term certainty) and other policy measures (such as subsidies and renewable energy standards) should promote such a shift.
- In its 20:20:20 package for Copenhagen, the EU has committed to 20% cuts below 1990 levels regardless of the outcome of international negotiations. If a satisfactory agreement on multilateral limitations commitments can be reached, the figure rises to 30%, paving the way towards an OECD-wide carbon market by 2015 and a wider global carbon market after that, with the CDM continuing to play a role but in a reformed, refocused, and more limited way.

- Even though the EU ETS provides the main source of demand for CERs, the CDM has fallen short in stimulating sufficient broad-base mitigation activity in developing countries. A stepping stone between the current state of affairs and assumption by major developing countries of a full cap-and-trade system is required. The EU has proposed adoption of sectoral crediting mechanisms to deliver both the necessary changes in actions and the requisite funding. The EU's de facto control over significant private-sector financial flows to the developing world through the offset crediting features of the EU ETS will likely enable it to secure adoption of this approach.

Experience to Date

Operating from 2005 to 2007, Phase I had a cap of 2.4 billion allowances per year. This first period was highly effective in making boardrooms aware of carbon risks and opportunities and stimulating the search for abatement opportunities. Further, the infrastructure of a functional, liquid market was successfully created. Phase I did, however, encounter problems. Allocations were not based on verified emissions. Also, companies were able to achieve steep initial reductions by exploiting cheap abatement opportunities previously overlooked. The result was a deep drop in allowance prices. The European Union (EU) responded by placing a firewall between Phases I and II by not allowing banking, thereby protecting Phase II from a flood of cheap allowances.

Currently the EU Emissions Trading System (EU ETS) is in Phase II (2008–12). The cap, significantly reduced from Phase I, is 2.08 billion allowances per year, 6.5% below verified emissions for 2005. The data received by the Commission indicate that there has been a sufficient drop in total EU emissions to make full compliance with Kyoto commitments appear likely. The use of certified emissions reductions (CER) and emission reduction units (ERU) offset credits from the Clean Development Mechanism (CDM) and Joint Implementation (JI) projects, authorized through the Linking Directive, has helped to lower compliance costs within the EU and generated significant levels of private investment in mitigation projects in a limited number of developing countries. The EU has placed a variety of both qualitative and quantitative limits on recognition of offset credits in order to protect the environmental integrity of the ETS from unsound credits or hot air credits, demonstrating the feasibility of both kinds of regulation. Because the EU provides most of the demand for CERs, the EU regulations demonstrate the potential for recipient cap-and-trade systems to profoundly affect the norms and practices for generating offset credits. The functional extension of the EU ETS to the countries of the European Economic Area, the linkage of the EU ETS to the Kyoto offset credit mechanisms, and the related switching of central registries from the EU-based Community Independent Transaction Log to the UN-based International Transaction Log are promising indications of the prospects for future linking of carbon markets.

One major lesson from both phases is that the initial allowance distribution process, based on grandfathered free allocation, is cumbersome, and a National Allocation Plan (NAP) system, under which allowance allocation was delegated to member states, fails to ensure the environmental integrity of the resulting cap. There were very different allocation methods in different member states' NAPs, leading to different treatment of similar types of industry. Auctioning of allowances by member states has been extremely limited (around 4%). These problems have spurred corrective measures in the design of the post-2012 ETS.

Reform

The EU ETS has been established as the core of EU climate policy and will continue in that role, although it currently covers only 41% of EU greenhouse gas (GHG) emissions; significant emitting sectors—including transport, aviation, and agriculture—are not subject to the ETS, being regulated by other policies and measures, but are being considered for inclusion at a later date. Legislation has already been put in place for the 2020 objectives, which include a goal to reduce economy-wide GHG emissions by 20% below 1990 levels regardless of what other jurisdictions do. The target will be raised to a 30% reduction if a new global framework can be agreed upon, although the difficult issue of how the burden of additional reductions will be allocated among the member states has yet to be agreed upon.

At the core of Phase III is a single EU-wide cap rather than independent caps for member states, thus abandoning the NAP system in favor of a centrally administered approach. Phase III requires emissions from sectors covered by the ETS to be reduced 21% below 2005 emissions. Domestic offsets are also being considered, along with coverage of additional gases and more industrial sectors, to bring nearly 50% of EU GHG emissions within the ETS. Also important is that the new legislation sets a policy of continuous reductions of 1.74% per annum beyond the end of Phase III in an effort to provide a degree of long-term regulatory certainty that has been lacking. Phase III requires a fully harmonized allocation process in which auctioning is the default allocation method: by 2013 more than 50% auctioning is anticipated, rising to 70% by 2020 and 100% by 2027. Member states must also use 50% of auction revenue for "climate purposes." A novel feature will be the ability to auction allowances early or auction more new entrant allowances to existing entities if the price spikes significantly in a short space of time.

Another significant feature of Phase III is that it provides a guaranteed level of demand for CDM and JI credits post-2012, although greater scrutiny of offset quality is to be expected, and quantitative limits will ensure that the Kyoto requirement of supplementarity will be preserved.

The difficulty in creating a sufficiently high price to incentivize domestic industry to develop and adopt more expensive forms of technology-intensive abatement such as carbon capture and storage (CCS) has been noted. Not all of this is necessarily due to the cap-setting and allocation decisions: it may also be due to the current economic climate and to lack of predictability on long-term policy, although it is difficult to distill the driving factors. It is hoped that consensus on international mitigation commitments can be reached in Copenhagen so as to trigger the 30% cut in EU emissions, which will drive the price path further up and promote long-term regulatory certainty. Other policy instruments, such as the renewable energy mandates and direct subsidies from the economic stimulus package, have also been adopted to stimulate a shift to low-carbon investment, and steps are underway to double R&D funding for low-emissions technology by 2012 and quadruple it by 2020.

International Offset Use

The EU ETS has, to date, been the main source of demand for CDM and JI credits. The EU has welcomed the CDM's contributions in engaging developing countries in carbon markets and stimulating investors and entrepreneurs to actively explore opportunities for reductions. The CDM has also proven an important cost-containment mechanism for Annex I

Parties and regulated firms. There are, however, important weaknesses in the CDM. As a project-based mechanism, it involves cumbersome procedures and large transaction costs, and it significantly limits the scale of mitigating activities. There have also been difficulties in ensuring the environmental integrity of CERs because of problems in defining baselines and applying additionality criteria and questions about the reliability of monitoring, recording, and verification (MRV) arrangements. Because the CDM provides and the EU ETS uses offset credits for reductions on a 1:1 basis, even environmentally sound projects do not reduce overall global emissions but simply relocate them from developing to developed countries, while unsound projects may actually lead to emissions increases. The participation of only a limited number of developing countries is also problematic. There is thus a major need for change if the CDM is to play a key role at the necessary scale in a global carbon market.

The EU has continually stressed the need for reform of the CDM to ensure that only projects that deliver real and additional reductions are credited and to extend the mechanism's reach beyond the low-hanging fruits of the very cheapest abatement opportunities. The EU does not advocate a total halt of the CDM by 2013, believing that a reformed CDM can and should play an important role not only in least developed countries (LDCs) but also in some sectors in key developing countries not suited for application of sectoral crediting mechanisms. It does, however, envisage that it will impose additional regulatory restrictions on recognition of CERs for the EU ETS, although details will depend on Copenhagen.

The EU's Vision for the Future

Progress has been made on developing the EU's long-term climate goals: the 20:20:20 by 2020 unilateral pledge, the 2020 objective of 30% below 1990 levels for the developed countries as a group, a stated undertaking to work with developing countries to reduce their business-as-usual (BAU) emissions by 15–30%, the 2050 global and developed country objectives, and, crucially, the overall objective of limiting warming to 2°C, which informs all of the others.

The EU Copenhagen Communication also includes a vision for the future of climate markets. It calls for linking the domestic cap-and-trade systems that will be adopted in the coming years to create a global carbon market, with a linked EU and US cap-and-trade system as the nucleus. The EU wants to see a robust Organisation for Economic Co-operation and Development (OECD)–wide carbon market by 2015, while exploring options for extending this network to other economically advanced countries by 2020. The European Commission's membership in the International Carbon Action Partnership (ICAP), a forum to share experiences and knowledge with the goal of linking climate markets, is a first step in this direction.

To help achieve the EU's goal of adoption by all developing countries of low-carbon development strategies by 2011, an international registry has been proposed in which all mitigation and adaptation measures taken by developing countries are recorded and can be transparently assessed. Developing country plans should have technical expertise to back them, which developed nations should help provide.

In addition to an improved CDM, there is a pressing need for a new mechanism to act as a stepping stone between project-based crediting and cap-and-trade in developing countries. The EU advocates the use of a sectoral offset crediting mechanism with baselines set below BAU, with the ultimate objective of phasing it out over time as participating developing countries adopt cap-and-trade systems that are linked to the global cap-and-trade allowance market. The sectoral mechanism envisaged would initially focus on the electric power sector and on sectors, such as aluminum and cement, that are subject to intense international competition. The EU has stated a preference for multilateral rather than unilateral criteria for such crediting mechanisms, which, it is hoped, will generate a substantial portion of the developing country mitigation investments required to achieve the 2°C goal. Other EU suggestions to raise the necessary funds from the developed world for developing country mitigation include payments into a central fund based on a formula that takes into account responsibility and ability to pay, and global allowance auctions.

FURTHER READING

A. Denny Ellerman, "The EU Emission Trading Scheme: A Prototype Global System?" *Harvard Project on International Climate Agreements* (August 2008).

European Commission, *Communication from the Commission to the European Parliament, the Council, the European Economic and Social Committee and the Committee of the Regions, "Towards a comprehensive climate change agreement in Copenhagen"* (January 2009).

European Commission, *Commission Staff Working Document, Accompanying Document to the Proposal for a Directive of the European Parliament and of the Council amending Directive 2003/87/EC so as to improve and extend the EU greenhouse gas emission allowance trading system, Impact Assessment* (2008), see (COM(2008) 16 final, SEC(2008) 53).

Chapter 25

Greenhouse Gas Emissions and Mitigation Measures in China

Jie Yu

Director, Policy and Research, The Climate Group (China)

Key Points

- Although China's GHG emissions account for 20% of the world's total, its per capita levels are still relatively low.
- China has already begun taking significant actions to mitigate future emissions growth, including adopting a goal of reducing emissions by 20% per unit of GDP by 2010.
- China views the transition to a low-carbon economy as an opportunity to develop valuable intellectual property rights and brands with global reach.

While China is currently responsible for 20% of global greenhouse gas (GHG) emissions, its per capita emissions levels are relatively low. As a result of its share of global emissions, industrialized nations have been pressuring China to adopt binding emissions caps. However, China has so far refused. Many may interpret China's reluctance to commit to a binding cap as a reluctance to confront the challenges of global warming. This is not the case. In fact, China is already heavily investing in major emissions reductions across a wide variety of sectors, in spite of the challenges these measures present to China's efforts to raise living standards for its population.

Furthermore, many in China have come to see the growing market for sustainable technology as an opportunity to diversify, further its economic growth, and reduce its foreign trade dependence (71% in 2007). Recently, Chinese companies have seen many successes in wind and solar energy, electric vehicles, and ultra-supercritical thermal power manufacturing. In light of these successes, many in the Chinese government believe that a global transition to more sustainable technologies will present an ideal opportunity for China to improve its research and development capacity, gain intellectual property rights for globally competitive technologies, and develop strong Chinese brands with global reach.

Basic Facts about China's GHGs Emission

The rapid economic growth and urbanization of China present both huge challenges and opportunities. In 2007, China's total emissions reached 760 million tons CO_2e, accounting for 20% of the world's total emissions. For the first time, China surpassed the United States as the largest emitter of greenhouse gases. However, its per capita emissions rate is only 4.3 tons CO_2, lower than the rate of all industrialized nations and far lower than the US's rate of 19.9 tons CO_2.

For a number of reasons, China's emissions are likely to increase, and will only be mitigated with the rapid deployment of carbon-neutral technology. First, China is undergoing a particularly energy-intensive period to meet the requirements of infrastructure constructions and improvements in its citizens' living conditions. Consequently, industrial emissions account for 70% of its total, as opposed to 18% in the US. However, this percentage is likely to decrease in the future.

Second, China essentially functions as the world's factory. Fully 20% of China's emissions originate in manufacturing and transport of goods for export. Such export emissions are likely to constitute a substantial portion of China's overall emissions for the foreseeable future.

Third, if China maintains 7.8% annual gross domestic product (GDP) growth, its business-as-usual (BAU) emissions will grow 3.1% annually. Under these assumptions, China's emissions growth will increase 113% from 2005 levels by 2030.

In light of these facts, transitioning to low-carbon technologies will benefit both China's energy security and global climate change mitigation efforts.

China's Domestic Mitigation Actions

In 2006, China launched its first national energy efficiency target: to reduce energy consumption across the economy by 20% per unit of GDP by 2010. This is part of China's eleventh five-year plan, and the target has been allocated to various sub-national governments.

In addition to the economy-wide target, China has designed and rolled out many more specific implementation programs over the years, including:

- *Industrial sector*: The Medium and Long Term Energy Conservation Plan contains medium- and long-term energy efficiency objectives for a dozen major industrial products, including steel, copper, and cement. It also includes targets for major energy-consuming equipment, such as coal-fired industrial boilers, medium- and small-sized motors, and specific industrial processes. This project will cost the public and private sectors more than USD 55 billion. It will save about 300 million tons coal equivalent and account for roughly 40% of the total reductions necessary to reach the national energy efficiency goal. The Top-1,000 Energy-Consuming Enterprises program is a central part of this effort. Under this program, energy efficiency targets were assigned for 1,000 major Chinese enterprises which collectively account for 47% of China's total industrial emissions. Key features of this plan include energy auditing and management institutions developed with the assistance of the government. The total anticipated savings are 100 million tons coal equivalent.
- *Power sector*: The power sector is responsible for 50% of the China's total emissions. Emissions will be limited by both increasing the capacity of renewable energy and improving the energy efficiency of the conventional power sector. Measures intended to increase efficiency of the traditional power sector include the replacement of small units with large ones to increase single-unit capacity; the development of cogeneration and related technologies; the promotion of large grid interconnection and efficient grid operation technology; and the replacement of small oil-fired generating units with units powered by natural gas. Additionally, 5% of the total power generation capacity was prematurely retired in 2006—mostly carbon-intensive plants. The national Ultra High Voltage Grid and Strong Smart Grid plans have also been rolled out recently to improve the

electricity transmission efficiency. From 2006 to 2007, the coal combustion efficiency increased 7% in thermal power sector due to these measures. China's Medium and Long Term Development Plan of Renewable Energy contains several specific targets intended to increase the importance of renewable energy: by 2010, the consumption of renewable energy will account for 10% of total consumption; by 2020, this proportion will increase to 15%.

- *Building sector*: Major policy goals include national design standards mandating 50% energy conservation for all newly constructed buildings, as well as more stringent standards mandating 65% energy conservation for new buildings in 4 municipalities and some other major cities, such as Beijing, Tianjin, Chengdu, and Chongqing. These measures should result in a 240 million ton carbon dioxide equivalent (CO_2e) reduction, which will account for 21% of China's entire energy conservation plan.
- *Transportation sector*: China also intends to increase the availability of public transportation and use of energy efficient vehicle technology. In the next three years, China plans to invest USD 500 billion on new railway construction. The Adjustment and Revitalization Plan of Automobile Industry, released by the State Council in March 2009, proposes that China increase the share of new energy vehicles and compact vehicles to about 5% of new auto sales from 2009 to 2011.

According to Chinese Development and Reform Committee's statistics, a successful reduction of national emissions by 20% per unit of GDP will result in a 750 million ton CO_2e reduction and conserve 300 million tons of coal equivalent. These reductions would be larger than the cumulative reductions made by Annex I countries under the Kyoto Protocol during the first commitment period.

The Perspective

China hopes that by rapidly scaling up the implementation of new technologies, it will rapidly reduce costs through economies of scale and development of new technology. Technology is the engine, policy the wheel, and finance the fuel. The hope is that by providing the right policy incentives, finance will flow to technology innovation and deployment. China has actively encouraged this process, even during the economic downturn. Although global demand for low-carbon technology declined, the Chinese government decided to grow the domestic market for low-carbon manufacturing to compensate for the shortfall in global demand.

One example of this strategy has been the implementation of ultra-supercritical power plants to replace older, less efficient power plants. From 2004 on, new plants that exceed 600 MW being brought online must use supercritical and ultra-supercritical thermal power technology.

Due to the scale of new power plant construction, the cost of those new technologies has dropped significantly. As a result, ultra-supercritical devices are able to compete with older, less efficient subcritical technology.

China has attempted to adopt a similar approach to the renewable energy market, including wind energy, solar energy, and electric vehicles. According to the industry association, the per unit cost of wind energy installation in China is now 30% lower than it was 3 years ago.

Conclusion

China understands the significant challenges that global warming presents and has initiated serious measures across wide sectors of its economy to address the problem. Although it has not committed to a binding cap, its GHG reductions could potentially, within a matter of only five years, match the reductions made by Annex I nations under the Kyoto Protocol. China further sees the spread of low-carbon technologies as an opportunity to diversify and strengthen its economy through the development of valuable intellectual property rights and brands.

FURTHER READING

On Chinese domestic mitigation actions and its monitoring, reporting, and verification (MRV) potentials, *see Mitigation Actions in China: Measurement, Reporting and Verification*, Working Paper of World Resource Institute, on behalf of Third Generation Environmentalism, available at http://pdf.wri.org/working_papers/china_mrv.pdf.

For more on Chinese domestic action with figures, see Julian Wong and Andrew Light, *China Begins Its Transition to a Clean-Energy Economy—China's Climate Progress by the Numbers* (Center for American Progress, 2009), available at http://www.americanprogress.org/issues/2009/06/china_energy_numbers.html.

For another summary of recent national mitigation policies, see Pew Center on Global Climate Change, *Climate Change Mitigation Measures in the People's Republic of China* (2007), available at http://www.pewclimate.org/docUploads/International%20Brief%20-%20China.pdf.

Chapter 26

Cities and GHG Emissions Reductions
An Opportunity We Cannot Afford to Miss

Partha Mukhopadhyay
Senior Research Fellow, Centre for Policy Research, New Delhi

Key Points

- Lower-carbon cities can substantially contribute towards mitigation efforts. Existing variations in energy use across cities have roots in local and national policies as well as patterns of behavior and cultural norms, all of which can be altered to reduce carbon intensity.
- Reducing carbon intensity of cities may not only require many conventional urban policies on financing and building codes to be re-examined, but also other macro policies such as tax breaks for homeownership and fiscal transfers to local government may need a fresh look. In particular, without changes in individual behavior, low-carbon cities are unlikely.
- Due to the rapid pace of urbanization and the immense lock-in effects once urban capital stock is built, policymakers may need to act even if the outcomes are uncertain. The wait for more clarity may be interminable and the consequences irredeemable.

Urban areas consumed about two-thirds of the world's energy in 2006. This is expected to increase to three-fourths by 2030. However, even in cities at similar levels of development, per capita urban energy use, and thus GHG emissions per capita, varies considerably. In light of this variation, would it be possible for governments to enact policies to promote less carbon-intensive cities? If so, what role could such policies play in a new climate change agreement?

An Underappreciated Opportunity

The average urban American consumes more than twice as much energy as the average urban European. Cities like Hong Kong, Tokyo, Singapore, and Amsterdam require less than a seventh of the energy of Houston, Phoenix, Detroit, and Denver to meet their transportation needs. Even within the United States, per capita energy consumption varies by a factor of three across cities. Many developing countries, especially India and China, are rapidly urbanizing, and similar discrepancies are beginning to emerge in these countries. For instance in China, energy use varies by a factor of seven from Chongqing to Hohhot, depending on income, climate, and energy intensity of industries. Given this variation, energy paths chosen by cities in emerging economies will have a huge impact on global GHG emission levels. In fact, lower-carbon cities could contribute over a third of the carbon mitigation in countries like India by 2050. This is an opportunity too big to miss.

Unfortunately, changes in city forms, behavior, and building types do not appear to be part of the mainstream climate change discussion. McKinsey's GHG cost curve, discussed extensively in this book, assumes very limited savings from behavior changes. The UNFCCC, in their "Investment and Financial Flows to Address Climate Change" in 2007, avers that "nearly all additional transport investment needed under the mitigation scenario is for the purchase of motor vehicles and production of transport fuels, [and] there will be no significant change to large transport infrastructure investments between the reference and mitigation scenarios." It also assumes that "most emission reductions in the buildings sector result from increased efficiency of appliances, space and water heating and cooling systems, and lighting."

What Opportunities Exist to Influence Energy Consumption in Cities?

Can policy actually make cities more compact, increase use of public transport, and affect building form? In order to encourage the development of more energy efficient cities, we first need to identify factors that may help explain variations in energy use across cities. While there is still debate, there seem to be some broad commonalities among cities with low energy use. Compactness of course helps, as residents travel less and use more public transportation. But building types, and the interaction of building type with behavior, seem to matter as well. In a survey in Taiwan, Hwang et al. (2009) found that 57% of respondents used the air-conditioner at work when they felt warm, but only 16% did so at home, while 58% used the fan or opened a window. While who paid the bill must have been relevant, it was also true that "only a quarter of workplaces . . . visited [were] equipped with fan or [had] . . . operable windows."

Nivola (1999) asked why European cities were more compact than American ones and offered the following answers: (a) less inner-city crime and (b) more investment in mass transit instead of highways, but also (c) agricultural support that raised land prices, (d) tax breaks for homeownership, (e) higher fuel and car taxes, (f) higher gas and electricity prices that make large homes and appliances expensive, and (g) higher share of transfers to local government. Thus, in addition to local policies, macro policies too appear to affect urban form, albeit in a complex and often poorly understood manner.

Even if these policies change, cities are limited in their response by the lock-in effect. This refers to the often-substantial impact of existing urban capital and systems on the cost of change. For example, Atlanta, where only 4.5% of the trips are by public transport, would need to increase its 74 km of track by 46 times and add 2,800 stations to get the same level of metro accessibility as similarly sized Barcelona, where 30% of trips use public transport, even though it has only 99 km of track and 136 stations. The lock-in effect has two major implications. First, proposals to change existing urban environments will be expensive. Second, the later one acts, the more new urban development will be locked into forms that are not compact and energy efficient.

Specific Policy Options

These complexities and lock-in effects make changing urban form and behavior a wicked problem, one that is almost incapable of resolution. A re-examination of some common urban policies below from a climate change perspective illustrates this difficulty of crafting solutions.

Property Taxes

Does the use of property taxes as the mainstay of local financing induce sprawl by discouraging densification since that leads to increase in taxable value or even by giving small groups the ability to choose their taxation levels by incorporating a new town? If so, an inter-governmental fiscal system that limits local taxes and relies more on statutory transfers to local governments and user fees may encourage more large, compact cities and fewer small towns.

Tax Benefits for Homeownership

Similarly, tax benefits for homeownership promote development of locations with low land values and hence home prices, usually at the fringes of the existing city. Disjunctions between home and work locations increase travel demand as homeownership deters relocation closer to work. Transport demand could fall if more people rent rather than own their homes. Increasing the supply of rental housing and making homeownership less aspirational could be critical to a low-carbon city.

LEED-Certified Modern Buildings

Moving from sprawl to aesthetics, is a modern building just glass, steel, and central air-conditioning? LEED-certified modern buildings are now visible in India and China, but do they reduce actual energy consumption? Newsham et al. (2009) find that, while, as a group, LEED buildings consumed less energy per unit area, up to a third of them used more energy than their conventional counterparts and higher levels of certification did not imply better energy efficiency.

Climate-Responsive Architecture and Behavioral Change

There are other approaches to modern building that challenge the conventional aesthetic imagination. Jiang Yi (2009) posits two philosophies of building design and use, viz.: 人定胜天 (Rén dìng sheng tian), i.e., the triumph of man over nature, vis-à-vis 天人合一 (Tian rén hé yi), i.e., the oneness of man and nature. Climate-responsive architecture, which leverages climatic resources to reduce use of energy for heating, cooling, and lighting of buildings, and part-time, part-space air-conditioning (even if by relatively inefficient equipment) fosters user tolerance for a wide range of indoor temperatures and may use much less energy than centrally air-conditioned spaces with more energy efficient equipment.

Culture and patterns of behavior are clearly important. The characterization of Europeans as people who wear sweaters indoors in winter and Americans as those who do the same in summer may be apocryphal, but it does point to behavior and culture as being critical elements. These differences may have roots in deeper cultural orientations. Is it possible that China and India, with distinct cultural sensibilities, will think and thus build differently than Western nations? Can their construction workforce, at the bottom of their labor totem pole, acquire the ability to erect such buildings?

Implementing Change: Governance and Ethical Concerns

Finally, who will make decisions about what policies to implement and how to finance them? Different layers of governance—international, national, and local—will all need to be involved in different capacities to influence urban energy consumption. For example, global agreements are needed to make international financial flows possible; action by national governments is required to change the tax structure, and only local governments are likely to be able to ensure building codes appropriate to their local environment.

These issues of multi-level governance are further complicated by matters of detail. If Annex I countries do decide to finance more efficient city building in developing countries, how should these transfers be structured? Approaches centered on crediting, which tend to rely on a form of BAU baseline or efficiency target, are unsuited for these kinds of systemic changes. Conditioning on GHG reduction would deny the uncertain and complex linkages between action and outcome. Instead, a specialized fund can support a set of measurable, reportable, and verifiable (MRV) climate-friendly actions in cities through the provision of long-term low-interest loans or interest-free, non-repayable financial transfers.

But even if this were acceptable, can parties agree on the kind of investments to support? Should public transport be rail or road based? Do gas pipelines qualify—because they encourage fuel switching in transport and facilitate load-center gas plants? What about water recycling to reduce energy use in transporting water? Does public rental housing, as in Hong Kong, and the additional cost for low-carbon cement qualify? Finally, should one country's taxpayer pay for cutting property taxes in another?

This paper also raises a broader concern: should international actors try to influence societal behavior in individual countries? Cultural relativism advocates caution in efforts to induce behavioral changes. However, without changes in behavior, low-carbon cities are unlikely. A second question is whether efforts to change behavior are preposterous. A good response to this is the anti-smoking campaign. This is, however, not a first-choice strategy for the OECD countries, as illustrated by their focus on energy efficiency and technological fixes to decarbonize their cities. Still, this does not change the fact that, regardless of how efficient Atlanta's cars become, its residents are likely to emit more carbon than Barcelona's.

Any attempt to reduce urban emissions is fraught with uncertainty. The choice before us is either to try to remake our cities, in spite of the uncertainty, or wait and hope that the uncertainty lessens. The risk is that the wait may be interminable and, worse, the consequences irredeemable.

FURTHER READING

Marilyn A. Brown, Frank Southworth, and Andrea Sarzynski, *Shrinking the Carbon Footprint of Metropolitan America Washington* (Brookings Institution, May 2008).

Shobhakar Dhakal, "Urban energy use and carbon emissions from cities in China and policy implications," *Energy Policy* (in press, 2009).

Ruey-Lung Hwang, Ming-Jen Cheng, Tzu-Ping Lin, and Ming-Chin Ho, "Thermal perceptions, general adaptation methods and occupant's idea about the trade-off between thermal comfort and energy saving in hot-humid regions," *Building and Environment, Vol. 44, Issue 6* (June 2009).

Guy R. Newsham, Sandra Mancini, and Benjamin J. Birt, "Do LEED-certified buildings save energy? Yes, but . . ." *Energy and Buildings, Vol. 41, Issue 8* (August 2009).

Pietro Nivola, *Laws of the Landscape: How Policies Shape Cities in Europe and America* (Brookings Institution Press, 1999).

Horst Rittel and Melvin Webber, "Dilemmas in a general theory of planning," *Policy Sciences, Vol. 4* (1973).

The World Bank, *World Development Report 2009: Reshaping Economic Geography* (2009).

Xiu Yang, "Overall building energy data in seven Chinese cities," *Presentation to CCICED Task Force on Energy Efficiency and Urban Development* (February 2009).

Jiang Yi, "Harmonious with nature: The Chinese approach to building energy reduction," in Mark Kelly (ed.), *Public #5: A Human Thing* (2009), available at http://www.woodsbagot.com/en/Documents/Public_5_papers/Harmonious_with_nature.pdf.

Chapter 27

A Prototype for Strategy Change in Oil-Exporting MENA States?

The Masdar Initiative in Abu Dhabi

Sam Nader

Head, Carbon Finance Unit, Masdar

Key Points

- Masdar, supported by the Government of Abu Dhabi, is attempting to create viable renewable and clean energy solutions, to commercialize these solutions, and at the same time to create a culture of sustainable development in the MENA region.
- Masdar City, the world's first zero-carbon city, is one of the flagship projects, along with a 100 WM solar plant, an industrial hydrogen power plant, and a nationwide Carbon Capture and Storage (CCS) system.

The debate around climate change and energy security is by now well known. A key issue facing our world today is how to tackle these challenges in a way that can sustain human progress and economic development, while at the same time safeguarding our environment and the future of our planet. It is clear there is no single answer to these challenges. Rather, the solution lies in the diversification of technologies, including clean fossil fuel energy, as we transition towards a low-carbon future. It is with this in mind that Abu Dhabi launched the Masdar initiative in 2006, taking the lead in developing a new model for government and business to work together in turning the world's climate and energy challenges into opportunities for sustainable growth and economic development.

A wide-ranging, multifaceted initiative, Masdar integrates the full renewable and clean technology life-cycle—from research to commercial deployment—with the aim of creating viable alternative energy solutions in a nascent and often fragmented industry. Benefiting from the full support of the Government of Abu Dhabi, Masdar provides a platform for the development of renewable energy and low-carbon technologies at a global level while creating a new clean energy growth-generating sector in the Emirate. The initiative is driven by five key components: education and research, project development, technology funding, value chain industry, and sustainable living.

With much of the world's carbon emissions increasingly coming from power generation, Masdar's investment and project deployment strategy in Abu Dhabi is focused on deriving a considerable share of future power supply from clean energy sources. This will be achieved by leveraging two of the Emirate's great natural advantages: year-round sunshine to produce solar power, and the development of fossil-fuel-based clean power generation projects on the back of a long-established hydrocarbon sector.

Masdar has already launched a 100 MW concentrated solar power plant in Abu Dhabi which will be operational by early 2012. This will be followed by a series of similar projects combined with next-generation photovoltaic power plants in order to reach a target of 1.5 GW of solar electricity by 2020, out of a projected installed capacity of 20 GW.

However, with Abu Dhabi's rapid increase of electricity demand over the next decade, the reliance on fossil fuels will likely remain high. Masdar is working on making our dependence on fossil fuels more sustainable, by advancing and rolling out multiple clean power technologies including pre-combustion and post-combustion carbon capture solutions.

The development of a national CCS network by 2020 forms the backbone of this effort. This program consists of a series of CCS projects aimed at taking a significant cut from Abu Dhabi's carbon footprint by 2020. The Phase I project started in summer 2008 and will be completed in 2014. Once fully operational, the project will capture around 5 million tons of CO_2 per year—equivalent to removing over a million cars from the roads of the United Arab Emirates—from conventional gas-fired power plants and heavy industry, using chemical absorption technology. The CO_2 will be transported in a pipeline network for injection in Abu Dhabi onshore oil reservoirs.

Masdar has also launched the world's first industrial-scale hydrogen-based power plant. The 400 MW plant will separate natural gas into hydrogen and CO_2 through auto-thermal reforming. The hydrogen is then burnt to produce emissions-free electricity, while the CO_2 is captured and sent into the CCS pipeline network.

Another flagship project, and a very tangible manifestation of Masdar's vision, is Masdar City, the world's first zero-carbon, zero-waste, and zero-car community under construction at the outskirts of Abu Dhabi. The city —which upon completion will be home to 90,000 people—will be fully powered by renewable energy and will showcase advanced technology in energy efficiency and green building. It will consume around 200 MW of power, compared with 800 MW normally required by a conventional city of the same size.

Masdar City is a prototype demonstration of how clean technologies and energy efficiency solutions can be integrated to provide a healthy emission-free environment with a high quality of life. Many elements of Masdar City will serve as best-practice examples for the blueprints of new and existing cities. Masdar City will provide us with great opportunities and a new way of life: sustainable industries, green jobs, and new, clean sources of energy. It will also provide the world with a successful model of sustainable living.

We believe that all of these initiatives and projects will have a substantial and growing impact on Abu Dhabi over the coming decade in reducing emissions and developing human capital. Although Masdar is still young, it is already serving as a catalyst for change in the region and is rapidly developing into a global leader in the renewable and low-carbon space. At the same time, Masdar is laying the groundwork for a growing awareness of sustainable development in the Middle East.

Part V

Climate Finance and World Trade Organization (WTO) Law and Policy

Chapter 28

The WTO and Climate Finance

Overview of the Key Issues

Gabrielle Marceau

Counsellor, Office of the Director-General, WTO;
Professor, University of Geneva Law School

Key Points

- While the primary goal of the WTO is to prevent unjustified restrictions on trade, the WTO has shown sufficient institutional and normative flexibility to allow member states to address environmental concerns effectively; this should remain true with actions relating to climate change mitigation and adaptation.
- The WTO will play a central role in resolving tensions between WTO Members' domestic policies to limit emissions and their obligations under WTO rules.
- As the mechanisms currently open to the WTO to confront climate change are limited, the primary effort to mobilize mitigation must come from international agreements.

Climate change, being such a broad issue, intersects with a number of areas of World Trade Organization (WTO) work, although the WTO's primary focus is to fight distorting trade restrictions. It is often suggested that WTO rules will be in conflict with domestic actions taken under the United Nations Framework Convention on Climate Change (UNFCCC) or other similar multilateral environmental agreements (MEAs), but this need not be the case. The WTO, like the UNFCCC, strives to ensure sustainable development. This brief essay first outlines climate change issues within WTO law, including the Doha Development Agenda (DDA), and then addresses some of the areas of potential tension between specific climate mitigation actions and obligations under the WTO.

The WTO's core activities are to negotiate reductions of tariffs and subsidies; to prevent domestic regulatory and other measures that unjustifiably restrict trade; to monitor domestic actions that may affect trade; and to settle disputes among its members. Basic rules include, among others, (1) the prohibition of unjustifiable discrimination between imported and domestic like products and (2) the prohibition of unjustifiable border import and export quotas.

Though created following World War II to stimulate the global economy, the WTO has demonstrated an institutional and normative capacity to adapt to the changing needs of its members. Although WTO has not yet discussed or acted on climate change per se, it is inevitable that it will do so in the future. And while WTO jurisprudence has not yet responded to the needs of climate change, it has been responsive to other new environmental needs of members. Therefore, when the WTO deals with climate change, it will benefit from the clarifications of WTO law on the scope of the environmental exceptions in General Agreement on Tariffs and Trade (GATT) Article XX; WTO Appellate Body (AB) decisions have been used to clarify relevant terms, conditions, and issues; and Members, responding to societal changes, have adopted waivers and even amendments to basic WTO provisions. Finally, some Members are talking about a temporary dispute peace-clause for climate-related issues. This could allow Members to rapidly adapt their domestic regulatory systems in a WTO-consistent manner to the needs of climate change mitigation.

In this paper, the issue of climate change is addressed from the perspective of the existing provisions of the GATT and the environmental exception in GATT Article XX in particular, as well as how trade negotiation can also facilitate climate change mitigation and adaptation measures.

GATT Article XX

Article XX enumerates a list of general exceptions that allow Members to give priority to policies other than trade, such as the protection of the environment. Generally, in WTO law a government is entitled to set the level of environmental protection it considers appropriate. Article XX authorizes such environmental measures that may incidentally violate other WTO obligations if they are "apt to contribute materially to the policy goal at issue." Importantly, the contribution of the environmental measure to the policy goal does not need to be immediately observable. As the AB noted, "it may prove difficult to isolate the contribution to public health or environmental objectives of one specific measure from those attributable to the other measures that are part of the same comprehensive programme." This is very relevant as climate change is a global phenomenon, and the contribution of any single domestic climate change mitigation measure to global mitigation will be very difficult to establish.

An important unresolved issue is the extent to which the environment exception of Article XX can be invoked against violations of WTO provisions other than those of the GATT (for instance, to the provisions of the Subsidies and Countervailing Measures Agreement that can become relevant if governments issue free emissions allowance as part of a cap-and-trade regulatory program) and with what effects.

Doha Development Agenda (DDA)

In the ongoing DDA, Members are negotiating enhanced tariff reductions on "environmental goods and services" that should favor the trade of the most needed clean technologies. Currently, the US imposes tariffs (topping out at 5.2%) on 32 of the 43 climate-friendly technologies identified by the World Bank. China imposes duties on all but two of the product categories, with a maximum rate of 35%. These tariffs are an impediment to trade and hinder the spread and development of clean technologies.

In the DDA, Members are also negotiating how to operate the relationship between the WTO rules and the commercial obligations in MEAs, which could become relevant if a treaty related to climate change (CC) is adopted.

Further, concluding the DDA would further open markets in favor of developing countries' exports and reduce trade-distorting agriculture protections, thus enhancing the economic power of developing countries and providing them with more means to take CC-related actions of all kinds.

However, it is worth remembering that both the DDA and GATT Article XX were not designed to deal with climate change issues. Recognizing this, WTO Director-General Lamy insists that once a new multilateral agreement on climate change is adopted, the WTO will be able to act effectively to allow members to implementation their CC commitments harmoniously with their trade obligations.

Potential Tensions between Climate Policy and International Trade Law

With the increase in domestic climate change regulation, the potential for tension between it and Members' WTO obligation increases. This section lists some of these potential areas of tension.

National Treatment and Most-Favored-Nations Obligations

All domestic regulations and taxation systems that potentially affect trade are subject to the national treatment and most-favored-nation obligations of the WTO—a very broad and powerful set of obligations. This means that imported and domestic "like" products—defined as products that compete with each other—must be treated similarly. Thus, a product coming from a country where there is a climate change program and another product from a country where there is no such program are "like" to the extent that they compete with each other and therefore must be treated the same way, unless the Article XX exception is invoked to justify such violation.

But if the environment exception is invoked, the importing country's environmental measures must be "apt to contributing materially" to the policy goal invoked—that is alleviating climate change—and such measures must be implemented in good faith. This means that countries in the same conditions must be treated similarly, and the level of development of the exporting countries must be taken into account; in addition, according to WTO case law, specific climate change actions undertaken by specific exporters (distinct from their government actions) would also have to be taken into account. For example, following the Shrimp-Turtle AB decision, even though domestic regulation may allow imports only from a country that has a climate change mitigation program, it could be argued that it must allow imports from a non-complying country if specific exporters within that country take comparable climate change mitigation actions.

So-called border tax adjustments raise significant issues under the national treatment obligations; when can a WTO member impose at the border a tax or a tariff against goods coming from a country that may not have a climate change program? When can a member offer its producers a tax rebate on their exports? What is the use of Article XX when environmental leakage is invoked to justify a violation of WTO rules?

Agreement on Technical Barriers to Trade (TBT)

As a result of the TBT, the WTO has rules applicable to domestic standards regulating products, the preparation and application of those standards, and their mutual recognition. For instance, government standards on logging certification and other forest product regulations adopted by Members as part of their responses to climate change must respect the prescriptions of the WTO TBT Agreement. The same is true for all energy efficiency standards, electricity standards, eco labels, certification schemes, etc.

Another important rule of the WTO (mentioned in the TBT and Sanitary and Phytosanitary (SPS) Agreements) is that if a domestic regulation complies with an existing international standard, such domestic regulation is presumed to be WTO consistent even if it restricts trade. At the moment, no such international standards relating to climate regulation exists. However, if specific climate regulatory standards were negotiated internationally, it could be argued that a domestic regulation implementing such standards could benefit from the WTO presumption of compatibility.

Also relevant is how to deal with private standards, such as those established by industry groups, NGOs, or the International Standards Organization. Such standards are generally not subject to WTO disciplines, but may become so if they are sponsored or promoted by Members. WTO law is not clear on this question.

Free Emissions Allowances

The WTO has rules concerning the level of specific production subsidies that will be allowed; such subsidies are restricted when they cause adverse effects on trade and international competition. Additionally, there are prohibitions on export subsidies. These provisions are relevant to domestic GHG emissions trading schemes that issue free allowances to local producers. As well, the WTO Subsidy Agreement and the national treatment allow for some forms of export tax-product rebates, subject to certain conditions. However, an economy- or sector-wide tax (as would be likely under a climate change program) is not product-specific, and so unable to be rebated upon export. Finally, the border administration of licenses will also be subject to the requirement of the Import Licensing Agreement, with regard to notification, transparency, and other administrative issues.

Agriculture

WTO rules to reduce distorting subsidies in agriculture can also become relevant for the protection of the environment. Reducing distorting subsidies will tend to favor the more naturally efficient agriculture producers and thus reduce the overuse and environmental abuse of agricultural land, which can result in high GHG emissions. On the other hand, the Agreement on Agriculture provides for unlimited "green subsidies," the full potential of which needs to be explored for climate change programs. Agriculture, which is one of the sectors most vulnerable to climate change, is also a key sector for international trade through subsidies to bad fertilizers, bad feedstock for animals, subsides to dedicated energy crops to replace fossil fuel use, improved energy efficiency, etc.

Regional Trade Agreements (RTAs)

As it is not clear when international agreement might be reached, it is quite possible that members of regional trade agreements will negotiate CO_2 standards, or even climate change conditioned rules of origin. As the WTO has rules on RTAs, the question arises as to how it should reconcile climate change actions taken by Members on the national, regional, and multilateral levels.

General Agreement on Trade in Services (GATS)

The rules on trade in services could also become relevant as they prohibit discrimination between foreign and domestic service providers. Trade of emissions allowances and other climate assets might be covered under GATS, and considered as of the same nature as "financial services." The GATS rules on investment (mode 3) may also become relevant as investment and competition-related actions will crucial to stimulate climate change mitigation programs.

Technology Transfer and the Agreement on Trade-Related Aspects of Intellectual Property Rights (TRIPS)

Finally, TRIPS rules are also very relevant. Mitigating climate change will be a major technological challenge. Of crucial importance will be technology transfer between countries; commercialization of low-cost technologies (many of which exist, but will need to be scaled up); and relations between innovation, patents, and compulsory licenses.

Conclusion

There is a significant overlap between climate issues and areas of WTO competence. As such, WTO rules should be kept in mind when constructing a post-2012 regime for climate change mitigation. Any new agreement need not be in conflict with the WTO.

FURTHER READING

A. Cosbey and R. Tarasosfky, *Climate Change, Competitiveness and Trade* (Chatham House Report, 2007).

T. Houser, R. Bradley, B. Childs, J. Werksman, and R. Heilmayr, *Levelling the Carbon Playing Field: International Competition and US Climate Policy Design* (Peterson Institute for International Economics and World Resources Institute, 2008).

G. C. Hufbauer, S. Charnovitz, and J. Kim, *Global Warming and the World Trade Organization* (Peterson Institute for International Economics, 2009).

G. Marceau and M. Cossy, "Institutional Challenges to Enhance Policy Coordination—How WTO Rules Could Be Utilized to Meet Climate Objectives?" *Proceedings of the World Trade Forum 2007, International Trade on a Warming Globe: The Role of the WTO in the Climate Change Debate* (2009).

J. Pauwelyn, *A Carbon Levy on Imports to Fight Climate Change* (Telos, September 2007), available at http://www.teloseu.com/en/article/a_carbon_levy_on_imports_to_fight_climate_change.

L. Tamiotti and V. Kulaçoğlu, "National Climate Change Mitigation Measures and Their Implications for the Multilateral Trading System: Key Findings of the WTO/UNEP Report on Trade and Climate Change," *Journal of World Trade* 43(5) (2009).

Chapter 29

Carbon Trading and the CDM in WTO Law

Robert Howse

Lloyd C. Nelson Professor of International Law, NYU School of Law

Antonia Eliason

Associate, Allen & Overy LLP

Key Points

- WTO rules are likely to play a central role in the regulation of carbon trading and other forms of carbon finance, both in the interim as climate finance regulatory bodies begin to address domestic measures affecting trading and in the long term as the carbon market becomes truly global.
- This paper examines some key issues in the evolving legal framework for international carbon trading and associated services, including the likely treatment under existing WTO agreements of the three Kyoto flexibility mechanisms and other trading systems for carbon assets.
- Although no policy exhortations are made here, it is clear that decisions about which legal provisions will regulate carbon finance will involve many complexities and have significant consequences, and therefore must be thought through carefully.

Capped Emissions Trading

The Kyoto Protocol authorizes three flexibility mechanisms to reduce the cost of compliance with its emissions targets. The first to be considered of these is a system of emissions trading among Annex I nations provided under Article 17, where countries with caps (calculated in assigned amount units, or AAUs) can reallocate the burden of abatement between them. Although the Protocol contains some general language regarding this system, including a requirement that Annex I Parties "strive to implement policies and measures . . . in such a way as to minimize adverse effects . . . on international trade . . . [and] on other Parties, especially developing country Parties," it provides very little specific guidance on the details of regulating international emissions trading, nor has significant progress been made in clarifying these arrangements. Given this absence, World Trade Organization (WTO) rules are likely to form a significant part of the relevant multilateral legal regulation.

One point to make clear is that trading of AAUs between states is governed by the Convention and Kyoto, whereas transnational transfers of permits recognized under domestic law as valid within domestic emissions trading schemes (such as the European Union Emissions Trading Scheme (EU ETS)) are not addressed by any international agreement. As yet, the WTO has not made a determination of whether and how any type of carbon market and the assets being traded falls under its auspices. Assuming the WTO would have regulatory jurisdiction, would these items be treated as financial services under the General Agreement on Tariffs and Trade (GATS) or as falling under some other GATS sectoral classification (perhaps environmental or energy services)? Alternatively, could they be considered goods under GATT, considering the carbon market primarily in terms of how it affects the terms and conditions of production of the goods for which carbon-based energy is an input?

While Article 17 authorizes emissions trading of AAUs only among states, it envisages that correlative carbon permits issued by states can be bought and sold directly between private parties or indirectly through brokers and exchanges. In practice, carbon trading seems very much like a financial service: the exchange of funds for an intangible right (to pollute). Moreover, there is no physical object that ever changes hands. That said, in their treatment by market participants, carbon permits also appear to be very similar to other basic commodities such as oil or corn, and these commodities are unquestionably goods. The answer may not be of an either/or character: as the Appellate Body held in *EC-Bananas*, the same regulatory scheme may affect trade in both goods and services, and therefore both the disciplines of the covered agreements on trade in goods and those of GATS may be applicable. Moreover, it is highly likely that regardless of the treatment of the underlying asset (i.e., allowances or credits) any financial products used within the context of carbon markets (e.g., derivatives such as swaps, futures, and options) will be treated as financial products and not goods.

If carbon trading is considered to be a financial service, then it would fall under the Annex on Financial Services to GATS. Finding carbon markets to be financial services under GATS would allow governments some latitude in taking prudential regulatory and other measures to protect their national markets and the international carbon market. Article 2 of the Financial Services Annex states that "a Member shall not be prevented from taking measures for prudential reasons, including for the protection of investors, depositors, policy holders or persons to whom a fiduciary duty is owed by a financial service supplier, or to ensure the integrity and stability of the financial system." If the "integrity and stability" of the carbon market is challenged, as may happen if allowances from other countries with emissions in excess of their caps are traded, the broad language of the Financial Services Annex will enable governments to support the market by excluding such permits if they do not conform to acceptable criteria.

Carbon trading also implicates the Subsidies and Countervailing Measures (SCM) Agreement. The definition of subsidy contained in Article 1.1(a)(1)(ii) of the SCM Agreement includes financial contributions "where government revenue otherwise due is foregone or not collected (e.g., fiscal incentives such as tax credits)." Article 1.1(b) lays out the other criterion for a subsidy—that a benefit be conferred by the financial contribution in question. Thus, if under any carbon trading system governments provide free carbon allowances that are then resold on the carbon market for a windfall profit, this may be viewed as a subsidy.

CDM and JI

The Clean Development Mechanism (CDM) and Joint Implementation (JI) are the other two Kyoto flexibility mechanisms, provided in Articles 12 and 6 of the Protocol, respectively. They achieve cost reductions by allowing developed countries to fund, directly or indirectly, emission reduction projects in developing countries (for CDM) or Annex I developed countries (for JI) and use the resulting certified emission reductions (CERs from CDM projects and ERUs from JI) towards meeting their own targets. Since these projects involve financing transfers to other countries as well as the possibility of technology transfer, a variety of WTO provisions are implicated. All of the relevant foregoing analysis from emissions trading could theoretically be applied to these mechanisms.

One way to consider these arrangements is as a transfer of emissions reductions between countries. Conceptualized this way, these projects could be seen as falling under GATS, as the trade in emissions reductions could be seen as trade in services: for instance, if a steel mill in Germany buys CERs from a wind farm in Morocco, this could be seen as the steel mill paying the wind farm to reduce the total GHG emissions of the two countries by a certain amount, with the CER as a mere certification of this service. This implicates most-favored-nation (MFN) provisions as well as National Treatment and Market Access provisions where a Member has bound the relevant sector(s) in its schedule.

Additionally, the WTO Agreements pertaining to trade in goods may apply (as suggested for international AAU/permit trading above) where the scenario above is rephrased in terms of the CERs being goods produced in Morocco and sold to a buyer in Germany or where inputs in energy production are concerned, for instance. The investment-oriented nature of these projects may also implicate the Agreement on Trade-Related Investment Measures (TRIMS). In the event that a project is inconsistent with either national treatment (GATT Article III) or quantitative restrictions (GATT Article XI), it would be in violation of TRIMS Article 2.1.

Two other potentially relevant WTO agreements are the Agreement on Government Procurement, since these projects involve cross-border investments under the supervision of governmental authorities, and the Technical Barriers to Trade (TBT) Agreement, which may apply where an Annex B country investing in a CDM project faces local technical regulations or conformity assessment procedures relating to products originating in the Annex B country.

It is quite likely that additional emissions credit offset trading systems between developed and developing countries will be established in connection with domestic ETS, such as the EU ETS and the US ETS provided by the Waxman-Markey legislation. In addition, arrangements to link domestic cap-and-trade systems will generate international emissions trading in allowances. These systems, arising initially under domestic law and agreements among specific states, will generate similar regulatory issues under international trade law.

RECs

Carbon trading is not the only form of instrument addressing greenhouse gas emissions. Whereas emission trading schemes involve the sale and purchase of entitlements to produce greenhouse gases, renewable energy certificates (RECs) serve to meet the requirement that a minimum share of electricity generated must come from renewable energy sources. Transactions in RECs are akin to emission trading schemes, but trade in RECs falls even more squarely into the realm of financial services, with the certificates usually being decoupled from the underlying energy being generated.

The analysis applicable to emissions trading above would also apply to trading of RECs, but due to RECs being decoupled from the actual energy being produced, provisions of GATS relating to transparency and disclosure, such as Article VI if licensing is required or paragraph 2(a) of the Financial Services Annex, will be particularly relevant to trade in RECs in order to avoid problems of accountability.

Conclusion

Because the United Nations Framework Convention on Climate Change/Kyoto regime has not resolved the regulatory uncertainties surrounding trading of AAUs/permits and project-based credit offsets, there is room for the WTO to play a role in providing additional regulatory support. WTO rules will also be highly relevant for new international offset credit and permit trading systems established pursuant to domestic law and agreements between individual states, and to international trading of RECs. The basis of WTO regulation could be found in existing, yet rarely used, agreements such as TRIMS and the Annex on Financial Services, as well as more frequently applied agreements such as GATS and the SCM Agreement. That said, the regulatory void surrounding international carbon trading highlights the need for an immediate solution with enforcement or adjudicatory capabilities, particularly in the current financial climate. The WTO can certainly help to fill the gap, but international climate regulatory laws and authorities must also address the issues. This is a priority for Copenhagen and beyond.

Chapter 30

Countervailing Duties and Subsidies for Climate Mitigation

What Is, and What Is Not, WTO-Compatible?

Robert Howse

Lloyd C. Nelson Professor of International Law, NYU School of Law

Antonia Eliason

Associate, Allen & Overy LLP

Key Points

- Subsidies are regulated by the WTO through the Subsidies and Countervailing Measures Agreement, which lays down rules for which subsidies are not permitted and recourse if they are used.
- One possible argument is that a state's omission to internalize the negative externality of climate change through domestic regulation can count as a subsidy, although the viability of this line of reasoning has been called into question.
- The allocation of free allowances to protect domestic industry from the competitiveness concerns of leakage raises subsidy issues, possibly even contravening WTO rules, and the same applies to certain efforts to promote renewable energy use.

Background on Subsidies in the Climate Change Field

The United Nations Framework Convention and Climate Change (UNFCCC) and the Kyoto Protocol adopt an approach to mitigation of climate change based on states binding themselves to reduce greenhouse gas (GHG) emissions to agreed levels, based on the notion of "common but differentiated responsibilities" for developed and developing countries. The Kyoto Protocol, however, does not specify the policies that states must use to achieve the bound emissions reductions, or the relevant desirability of different policy instruments. The Protocol merely provides a list of policies that states may use to achieve emissions reductions.

Many of these policies can be pursued either by regulatory measures —emissions caps, renewable energy mandates, etc.—and/or through subsidies that provide incentives to market actors to engage in behavior that leads, either in the short term or long term, to lower emissions. The Intergovernmental Panel on Climate Change (IPCC), in its Fourth Assessment Report, notes, "direct and indirect subsidies can be important policy instruments, but they have strong market implications and may increase or decrease emissions, depending on their nature. Subsidies aimed at reducing emissions can take on different forms, ranging from support for research and development (R&D), investment tax credit, and price supports (such as feed-in tariffs for renewable electricity)."[1] The International Energy Agency (IEA) in its database "Addressing Climate Change: Policies and Measures" distinguishes a range of policies that would be considered to have subsidy elements, at least from the perspective of international trade rules, including incentives/subsidies (direct payments to market actors); public investment; and research and development. The IEA database divides Climate Change Policies and Measures into those that support renewable energy and those that support energy efficiency. As is evident from an examination of the measures inventoried in the database, a wide range of IEA members and other states have implemented a variety of policies with elements of subsidies. The pervasiveness and diversity of such policies as means of implementing Kyoto obligations lead to important consequences both for global governance of climate change *and* for the international trading system, especially the World Trade Organization (WTO).

Subsidy Regulation under the WTO

The Uruguay Round Subsidies and Countervailing Measures Agreement (SCM) placed in the category of "prohibited" in the SCM Agreement export subsidies (subsidies given only for products that are exported) and domestic content requirements (requirements that goods sold in a country contain a certain minimum of domestic value added). The Agreement introduced a category of domestic subsidies called "actionable," which can be challenged in WTO dispute settlement proceedings, thus providing a multilateral legal remedy against subsidization. In order for a subsidy to be challenged in WTO dispute settlement as "prohibited" or "actionable," it has to fall within the definition of subsidy in Article 1 of the SCM Agreement, which means it must entail a "financial contribution" of governmental financial assistance to firms (from cash payments to equity infusions to provision of goods and services below market prices), and also confer a "benefit" on an enterprise; the subsidy must also be "specific," either de jure (legally targeted at a particular industry or enterprise or group of industries or enterprises) or de facto (in fact used only or disproportionately by a particular industry or enterprise or group of industries or enterprises). 2.1(b) of the SCM Agreement refines the concept of specificity:

> Where the granting authority, or the legislation pursuant to which the granting authority operates, establishes objective criteria or conditions governing the eligibility for, and the amount of, a subsidy, specificity shall not exist, provided that the eligibility is automatic and that such criteria and conditions are strictly adhered to. The criteria or conditions must be clearly spelled out in law, regulation, or other official document, so as to be capable of verification.

In the case of prohibited subsidies (i.e., export subsidies), specificity is presumed and does not have to be proven by the claimant.

If a subsidy meets the above criteria for actionability, a WTO Member may either challenge the subsidy in WTO dispute settlement, seeking the remedy of removal of the offending measure, or it may countervail the subsidy. If a Member pursues the first option, it must show the existence of certain "adverse effects" on WTO Members other than the subsidizing Member, including itself. These adverse effects are listed in Article 5 of the SCM Agreement, and include injury to domestic producers of a like product in competition with the imported subsidized product (injury in this sense must exist if countervailing duties are to be imposed); nullification or impairment of benefits accruing "directly or indirectly" under the

GATT, in particular tariff concessions; or serious prejudice to the interests of another Member. "Serious prejudice" is further defined in Article 6.3. To show "serious prejudice" the complaining WTO Member must show that the effect of the subsidy is to displace imports of a "like" product into the market of the subsidizing Member; or to displace exports of the complaining Member to a third country market; or significant price suppression or price undercutting in the same market with respect to like products; or finally "the effect of the subsidy is an increase in the world market share of the subsidizing Member in a particular subsidized primary product or commodity as compared to the average share it had during the previous period of three years and this increase follows as a consistent trend over a period when subsidies have been granted."

Where the Member chooses the option of imposing a countervailing duty (CVD), it must comply with the various procedural and substantive criteria in the SCM Agreement that apply in the case of CVD actions, including the requirement of showing "material injury." These criteria apply also where a Member is countervailing a "prohibited" subsidy. The SCM Agreement (Article 8) originally entailed a defined list of subsidies to be *deemed* "non-actionable," i.e., subsidies immunized from challenge in WTO dispute settlement as well as countervailing duty action, even if they were to be found to meet the criteria discussed above. This list included certain subsidies for research and development, environmental protection, and to disadvantaged regions. However, this provision for deemed non-actionability applied provisionally, for only the first five years that the SCM Agreement was in force. Since its effective expiration, WTO Members have been unable to agree to either continue with the list as it now stands or to create a different list. Therefore, today there are no subsidy programs that are explicitly protected as non-actionable.

Omission to Regulate—a Subsidy?

Joseph Stiglitz has suggested that the failure especially of the WTO Members not participating in the Kyoto Protocol to internalize the climate change costs caused by carbon emissions from the production of products is a "subsidy" to the producers of such products, resulting in a distortion of international markets in the trade in goods. Most WTO legal experts who have commented on Stiglitz's proposal have dismissed it as clearly not justified under the WTO rules in the SCM Agreement, since one or another of these criteria is obviously not met. According to Bhagwati and Mavroidis, "a subsidy exists only if a government has made a financial contribution or has incurred a cost. . . . The argument that the United States policy [of not participating in Kyoto] is a 'hidden subsidy' is irrelevant and cannot justify an EU action under the SCM Agreement."[2]

Nevertheless, among the meanings of "financial contribution" in the SCM Agreement is the government provision of goods or services other than general infrastructure. There are no pre-assigned property rights to the atmosphere; instead, states are generally thought to have prescriptive jurisdiction over this commons, subject to international obligations by treaty (e.g., the Kyoto Protocol) or custom. Thus, where a firm is allowed to emit carbon into the atmosphere up to a certain ceiling, this is not a consequence of some preexisting property right in the atmosphere that is being exercised by the firm, but rather, of the assignment of such a right or entitlement by the state to the firm in question. Such a right or entitlement is a valuable asset, indeed an asset that can be bought and sold in the marketplace. The question arises as to whether the failure to charge a market price for the asset in question constitutes the provision of goods or services, and therefore a financial contribution within the meaning of Article 1 of the SCM Agreement.

Leakage

Various policy measures have been proposed to address the problem of "carbon leakage"—the notion that where a jurisdiction imposes emissions caps on its industries, these industries may become uncompetitive relative to those operating in jurisdictions where no such caps exist, or lesser burdens to limit or reduce emissions. Both an increase in emissions caps and the provision of free allowances to selected industries would raise issues under the SCM disciplines. Since rights to pollute constitute provision of a valuable good by the government (access to an exhaustible natural resource), and thus a "financial contribution," whether these are provided in the form of basic entitlements up to a certain level, or as free allowances, they may well be actionable subsidies where they are specific (i.e., targeted at particular industries facing competitiveness pressures) or de facto (i.e., disproportionately or predominantly used by certain sectors).

Promoting Low-Carbon Investment

A wide range of subsidy programs purports to address climate change through reducing the cost of producing and/or consuming energy from non-carbon-emitting sources, relative to conventional, carbon-emitting energy sources. According to the IPCC in its Fourth Assessment Report, "One of the most effective incentives for fostering GHG reductions are the price supports associated with the production of renewable energy, which tend to be set at attractive levels. These price supports have resulted in the significant expansion of the renewable energy sector in OECD countries due to the requirement that electric power producers purchase such electricity at favorable prices."

In the *PreussenElektra* case, the European Court held that minimum-price purchase requirements under German law could not be considered "state aid" in European law because of the absence of any direct or indirect transfer of state resources.[3] In the WTO SCM Agreement, by contrast, a "financial contribution" includes a situation where "a government makes payments to a funding mechanism, or entrusts or directs a private body to carry out one or more of the type of functions illustrated in [SCM Agreement Article 1.1(a)(1)] (i) to (iii) . . . which would normally be vested in the government and the practice, in no real sense, differs from practices normally followed by government." Since SCM Agreement Article 1.1(a)(1)(iii) includes "purchasing goods," the argument is that a situation where the government directs a private actor to purchase goods at a higher than market price is included within the meaning of "financial contribution" even if the government does not incur any cost *itself*. In the *Canada-Aircraft* case (Paragraph 160), the Appellate Body observed that "financial contribution" could include those situations where a private body has been directed by the government to engage in one of the actions defined in the SCM Agreement Article 1.1(a)(1)(i)–(iii), even if the government does not bear the cost of such delegated action.

However, the German minimum-price purchase requirements do not necessarily constitute a "financial contribution" within the meaning of the SCM Agreement, because where the government entrusts or directs a private body, the SCM Agreement *also* requires that the function entrusted or delegated to the private body be one that is *normally* performed by the government.

In order to violate WTO rules, a subsidy has to have conferred a "benefit" on the recipient, i.e., a competitive advantage over and above general "market" conditions. Some programs for renewable energy may not confer a "benefit" in this sense. Measures that merely defray the cost of businesses acquiring renewable energy systems or which compensate enterprises for providing renewable energy in remote locations do not necessarily, for instance, confer a "benefit" on the recipient enterprise. They simply reimburse or compensate the enterprise for taking some action that it would otherwise not take, and the enterprise has not acquired any competitive advantage over other enterprises, which neither take the subsidy nor have to perform these actions.

With respect to the requirement of *specificity*, subsidies that are provided to *users* of renewable energy may well not be specific if they are available generally to enterprises in the economy.

FURTHER READING

International Energy Agency (IEA), *Addressing Climate Change: Policies and Measures Database,* available at http://www.iea.org/textbase/pm/?mode=cc.

NOTES

1. S. Gupta et al., "Policies, Instruments and Co-operative Arrangements," in *Climate Change 2007: Mitigation; Contribution of Working Group III to the Fourth Assessment Report of the Intergovernmental Panel on Climate Change,* B. Metz et al., eds. (Cambridge: Cambridge University Press, 2007), pp. 761–762.
2. J. Bhagwati and P. Mavroidis, "Is Action against US Exports for Failure to Sign Kyoto Protocol WTO-Legal?" *World Trade Review* (2007), 6: 2, 299–310.
3. Case C-379/98, *PreussenElektra AG v. Schleswag AG* (2001), I-2099.

Chapter 31

Border Climate Adjustment as Climate Policy

Alexandra Khrebtukova

Scholar, NYU School of Law Institute of International Law and Justice; Judicial Law Clerk, US Court of International Trade

Key Points

- Border Climate Adjustments (BCAs) are national measures based on the principle that climate costs should be imposed on GHG-intensive production at the point of market entry rather than the point of production.
- These measures impose a non-discriminatory price on imported GHG-intensive goods as a condition for market entry, complementing the imposition of climate costs on like domestic products via national regulation.
- Comporting with the destination principle, BCA measures may also be used to remit the costs imposed by domestic GHG-intensive goods regulation for goods destined for consumption and driven by demand from other markets, encouraging destination governments to similarly employ the market-access-conditioning approach to regulating GHG-intensive consumption.
- BCA measures can improve the political viability and environmental effectiveness of national regulation, and if the two are structured correctly, they can be permissible under the international trade law regime.

Distributing the global greenhouse gas (GHG) abatement effort (and the costs of that effort) necessary in light of Intergovernmental Panel on Climate Change (IPCC) findings is a daunting problem. It appears essential to regulate GHG emissions by putting a price, through a carbon tax or a cap-and-trade scheme, on tons of GHG emitted. Doing this through national regulation has the potential to cause "carbon leakage," shifting GHG-intensive production (such as iron, steel, aluminum, pulp and paper, and cement) towards jurisdictions with less stringent or no regulation. Globalized markets for these products make such shifts more possible, undercutting emissions control regimes.

Accordingly, measures to correct for the competitiveness-distorting/emissions-leakage effects of domestic GHG regulation may prove a necessary component of such national schemes, both as a matter of domestic political viability (to guard industry against unfair competition with foreign goods not subject to similarly stringent climate costs) and environmental effectiveness (to ensure that total GHG emissions are actually reduced). I will call such measures border climate adjustment (BCA) schemes, by which I will mean a general category of national regulations directed at certain categories of imported products, which seek to impose a total price on the production of these goods approximating the total price imposed on the production of like domestic goods.

The ultimate purpose of a BCA scheme is to substantially preempt emissions leakage. A BCA scheme (used in conjunction with a similar cost internalization scheme imposed on domestic producers supplying the national market) ensures that the domestic emissions reductions are not offset by the presence of non-regulated products in the marketplace. If GHGs emitted in the course of industrial production are regulated by a national cap-and-trade scheme coupled with a BCA for imports—that is, if the point of climate cost payment occurs at point of market entry in the destination market—the problem of emissions leakage does not arise to the same extent. Because foreign production costs are equalized with those of domestic production through BCA schemes, producers face the same costs of selling goods in the destination market irrespective of the level of GHG regulation in the country of origin.

The use of a BCA-enabled market-access-conditioning approach may facilitate the gradual build-up of an eventually comprehensive global GHG management regime by leaving it up to each state to effectively regulate its own contribution to ongoing GHG emissions from industrial production worldwide. To prevent against emissions leakage—that is, to effectively regulate some discrete portion of continued global GHG emissions which may be directly traced back to consumption demands within a given national market—State A regulates GHGs emitted in the course of producing only and all those units of (covered) production that enter its market, whether home-made or foreign. As products from State B incur costs when exported to State A (and so producers based in State B complain to their government), State B will seek in return to generate revenue from imposing its own climate costs upon goods imported from State A. Because World Trade Organization (WTO) Members are only permitted to impose costs upon imports from other Members evenhandedly with like costs imposed on like domestic products, the political feasibility of instituting domestic regulation in State B is thus increased.

Anticipating the likelihood that countries of origin significantly affected by BCA costs may seek to subject State A's exports to BCA as a condition for market entry, State A withdraws products destined for consumption in other markets from its regulatory scope, possibly through remitting allowances back to exporting producers. As States B, C, D, etc., begin to similarly regulate GHGs emitted because of consumption demands for certain GHG-intensive industrial production—that is, as other States begin to similarly condition access to their market (for both domestic and foreign covered goods) on the payment of a price for (approximately) each ton of GHG emitted per unit of production seeking market entry—an increasing quantity of GHG emissions attributable to global production effort will be placed within the scope of an effective (because not subject to emissions leakage) climate cost-internalization regime.

Because the regulatory purpose of a well-designed BCA, coupled with a national cap-and-trade scheme which initially allocates GHG permits by government auction, is essentially the same as that behind a direct tax levied at point of market entry for GHGs emitted in the course of certain products' production, such BCA may, in principle, be structurally conceived in the WTO as a legitimate border tax adjustment (BTA) scheme.

A Working Party established by the precursor to the WTO to analyze and clarify international trade law on BTAs adopted the following definition of taxes: "compulsory, unrequited payments to general government. They are unrequited in the sense that benefits provided by government to taxpayers are not normally in proportion to their payments."[1] The forced internalization of climate costs into costs of production through mandatory requirements to purchase and retire a number of GHG emission allowances or credits equal to the tons of GHG emitted in the course of a given compliance period easily fits within this broad definition. Leaving aside the special problems of allowances distributed to domestic industry at no cost by the government, the market price of GHG allowances paid to the government at auction, in addition to any penalties paid for every ton of GHG emitted in excess of surrendered allowances or credits, are payments to the government.

One could argue that a governmental program imposing a price on every ton of GHG emitted does not require unrequited payment because in return for payment, the regulated entity receives the right to pollute a quantity of GHG tons precisely in proportion to that paid for. Nevertheless, as a matter of public policy, GHG emission allowances should not be conceived as benefits in proportion to the payments made to the government in terms of their market price, as it would be inconsistent with the general spirit of national GHG-capping legislation to construe such an Act as creating beneficial rights to pollute when its long-term goals are in fact to drastically reduce or eliminate GHG emissions. Moreover, as prices increase over time (due to lower caps, higher taxes, or more stringent standards) the relationship between tax surrendered and "benefit" granted breaks down even further.

Importantly, the Agreement on Subsidies and Countervailing Measures (SCM) explicitly allows the remission of prior-stage cumulative indirect taxes on "energy, fuels, and oil used in the production process."[2] A WTO Member's domestic GHG management regime which mandates the payment of some price for every ton of GHG emitted in the course of GHG-intensive regulated entities' production effort over a given timeframe is essentially a scheme which imposes a tax upon GHG-intensive energy used in the course of certain industrial production: the majority of GHG tons emitted in the course of GHG-intensive production is due to the energy consumed in producing, rather than some other aspect of the production process. Accordingly, were a WTO Member to choose to regulate such GHG emissions on the destination principle—that is, to impose a price upon only those GHG tons attributable to products consumed on the home market—then, under the SCM Agreement, that Member could lawfully remit payment for such quantity of GHG that is proportionate to the portion of total regulated production effort that is exported to be consumed (and presumably regulated) in other markets.

The same legal principles that govern the adjustability of consumption taxes with respect to products destined for export also govern the adjustability for those same payments with respect to foreign products entering the home market for consumption. Because, as reported by the BTA Working Party, "GATT provisions on tax adjustment appl[y] the principle of destination identically to imports and exports,"[3] eligibility for adjustment with respect to the remission of taxes on exports destined for consumption in other markets ipso facto translates into eligibility for adjustment in the form of taxes levied on imports seeking access to the US market. Accordingly, prior-stage cumulative indirect taxes on GHG-intensive energy used in the course of production are equally adjustable with respect to imported products seeking access to a Member's market as they are with respect to products destined for consumption elsewhere.

Given that all BCA systems face the tough challenge of calculating the level of GHG embodied in imported products, I argue for the use of a BCA scheme based on the destination principle rather than the kind of measures included in many existing BCA proposals, which commonly use a "comparability-in-effect" test to establish whether imports come from a country with sufficient levels of GHG regulation. Calculating the comparability of other regulatory systems is notoriously difficult: a price on carbon can be used as comparator if a carbon tax or cap-and-trade scheme is used, but (i) price volatility, (ii) different system characteristics (e.g., coverage, offset use, intertemporal flexibility), and (iii) other regulation (e.g., renewable energy standards) make this comparison far from easy. Moreover, once a significant number of nations regulate GHG in a meaningful way, the administrative challenges faced by an agency tasked with performing these calculations will multiply exponentially. Regulation using the destination principle entirely avoids these issues and is more likely to be WTO-compliant.

In sum, room can and should be found in the global climate regime for more stringent unilateral action involving the use of non-discriminatory BCAs, which does not preclude the use of other measures to correct for historical responsibility or developmental inequities, such as side payments or technology transfer agreements. BCA measures, in conjunction with national cap-and-trade schemes which allocate capped tradable allowances by government auction, may not only be justified as a matter of world trade law, but may also offer unique benefits for the development of economically efficient and environmentally effective global GHG management. Conditioning market access for certain domestic and imported GHG-intensive goods on the purchase of GHG allowances for every GHG-ton emitted in the course of production may thus provide an important climate policy mechanism, encouraging the gradual establishment of a transborder administrative regime for coordinating the appropriate levels of cost distribution necessary to eventually steer the globe toward both a well-functioning climate and a well-functioning economy.

FURTHER READING

Javier de Cendra, "Can Emissions Trading Schemes Be Coupled with Border Tax Adjustments? An Analysis vis-à-vis WTO Law," *15 Review of European Community & International Environmental Law 131* (2006).

Paul Demaret and Raoul Stewardson, "Border Tax Adjustments under GATT and EC Law and General Implications for Environmental Taxes," *28 Journal of World Trade 5* (1994).

Roland Ismer and Karsten Neuhoff, "Border Tax Adjustment: A Feasible Way to Support Stringent Emission Trading," *24 European Journal of Law & Economics 137* (2007).

NOTES

1. *See* WTO, Committee on Trade and Environment, Note by the Secretariat, Taxes and Charges for Environmental Purposes—Border Tax Adjustment, WT/CTE/W/47, 2 May 1997, at 6 (citing BISD 18S/97).

2. World Trade Organization, Agreement on Subsidies and Countervailing Measures, Annex I, (h) and accompanying Notes. *See also id.*, Annex II n.61 (allowing remission of prior-stage cumulative indirect taxes on inputs "that are consumed in the production of the exported product (making normal allowances for waste)").

3. GATT, Border Tax Adjustment Working Party, at ¶10 (quoted in Paul Demaret & Raoul Stewardson, *Border Tax Adjustments under GATT and EC Law and General Implications for Environmental Taxes*, 28 J. World Trade 5, 30 (1994)).

Chapter 32

Enforcing Climate Rules with Trade Measures

Five Recommendations for Trade Policy Monitoring

Arunabha Ghosh

Oxford-Princeton Global Leaders Fellow, Woodrow Wilson School, Princeton University

Key Points

- Developing countries are rightly wary of pro-climate trade measures being used as protectionism by developed countries, and also about formulation of new trade rules and classifications for environmental services and embedded carbon in ways that favor developed country interests.
- Developing countries need to build greater capacity to monitor the trade policies of other countries, to detect in time and challenge disguised protectionism.
- The WTO Trade Policy Review Mechanism should be strengthened to combat environmental measures that might be protectionist.
- Developing countries need to increase their expertise and influence on climate-related services, standards, and labels, or the rules will become skewed against their interests.
- Emissions measurement and self-reporting capacity in developing countries must be greatly strengthened.

Laws being drafted or proposed in developed countries envisage the use of trade sanctions to induce participation by other countries in a global climate regime, or to level the playing field for businesses and avoid relocation and carbon leakage, or to punish non-compliant countries. It is possible that an international climate agreement may eventually authorize certain trade sanctions, as was done in the Montreal Protocol on the stratospheric ozone layer and for other environmental aims. New rules and definitions are being developed on issues such as liberalization of trade in environmental goods and services, and on specifications for measurement of embedded carbon and emissions, which may disadvantage developing countries. Several essays in this volume highlight different areas in which climate law and policy are already having to take account of World Trade Organization (WTO) agreements on trade and market regulation, including on trade restrictions, subsidies, taxes, and carbon labeling. Linkage between climate mitigation and trade law is inescapable, and offers both attractions and threats from the viewpoint of developing countries. This essay focuses first on major concerns developing countries have about such linkages, and then proposes five specific ways to ameliorate these concerns.

What Are Developing Countries' Concerns with Trade and Climate Linkages?

The primary motivation for using trade measures is the fear of industrial competition from non-participating countries. A secondary preoccupation is that emissions will increase elsewhere due to carbon leakage if firms relocate to countries with lower environmental standards. While the evidence for leakage and competitiveness threats is mixed—and restricted to a few sectors—proposals for linking the trade and climate regimes have gained momentum.

From the perspective of developing countries, any serious attempts or threats to affect trade through climate measures prompt a variety of concerns. Four sets of concerns about the legality and governance of such measures can be noted here.

First, protectionism may be disguised as climate-friendly policies. The incentive to exaggerate the extent of carbon leakage is strong, and special interests could hijack trade measures for protectionist purposes.

Second, although the WTO's Technical Barriers to Trade (TBT) Agreement governs standards and labeling, it does not apply to private businesses. Therefore, firm-led decisions to regulate emissions by introducing labeling requirements and standards could adversely affect exports without the protection of WTO rules.

Third, the relaxation of trade barriers against environmental goods and services (EGS) may be applied unevenly and disproportionately benefit developed countries. The liberalization of trade in EGS, which includes products and services that yield environmental benefits, such as catalytic converters and consultancy services on wastewater management, is part of the Doha Round of negotiations. The global market in EGS is estimated to be about USD 550 billion. Developing countries, on average, have low applied tariffs against EGS and view demands to reduce barriers as a strategy of rich countries to promote new industrial sectors.[1] Yet, when it comes to their export interests in energy-related goods, developing countries face trade barriers abroad. Brazil's dispute against a ban on ethanol exports to the United States or China facing anti-dumping duties against energy-saving light bulbs in the European Union (EU) for several years are cases in point.

Fourth, since developing countries demand technology transfer as a condition for reducing emissions, they have concerns about how stringently intellectual property rights (IPRs) are enforced by the trade regime. Stringent IPRs could increase the costs of technology, disadvantage firms in developing countries, and undermine domestic absorptive capacity for new technologies. Compulsory licensing, exemptions from patentability, forgoing patents on publicly funded research, and multilateral funds to buy out patents are means of facilitating technology transfer that developing countries might advocate in the trade and climate regimes.

Suggestions for Trade Policy Monitoring

Concerns about emissions leakage, industrial competitiveness, and market access cannot be resolved without confidence in the measurement, monitoring, and enforcement mechanisms in the trade and climate regimes and within all states involved. Compliance with negotiated rules is contingent on credible monitoring: states are likely to renege on commitments if they believe that their actions will not be easily detected or monitored. In light of the preceding discussion, here are five suggestions for strengthening trade monitoring and environmental measurements.

1. Recognize Capacity Challenges for Monitoring Trade Measures

A first line of defense against illegitimate trade measures is regular monitoring. Export-oriented firms could keep a lookout for policy changes abroad, but effective monitoring requires institutional capacity. Many countries collect commercial intelligence through trade attachés in embassies or via industry bodies. A more formalized process would include a dedicated state agency with the mandate for monitoring trade barriers. The most institutionalized approach at the country level involves regular publication of foreign trade barriers reports, which when disseminated widely give valuable information on existing and anticipated measures.

Few countries, however, have the kind of institutional capacity needed to monitor climate-related trade measures. A recent analysis of seventy developed, developing, and least developed countries (just under half the WTO's membership) found that only half of them collected commercial intelligence on a regular basis, and less than a fifth published regular reports on foreign trade barriers (Figure 32.1).[2] A few large developing countries have built capacity for monitoring specific areas (say, Brazil

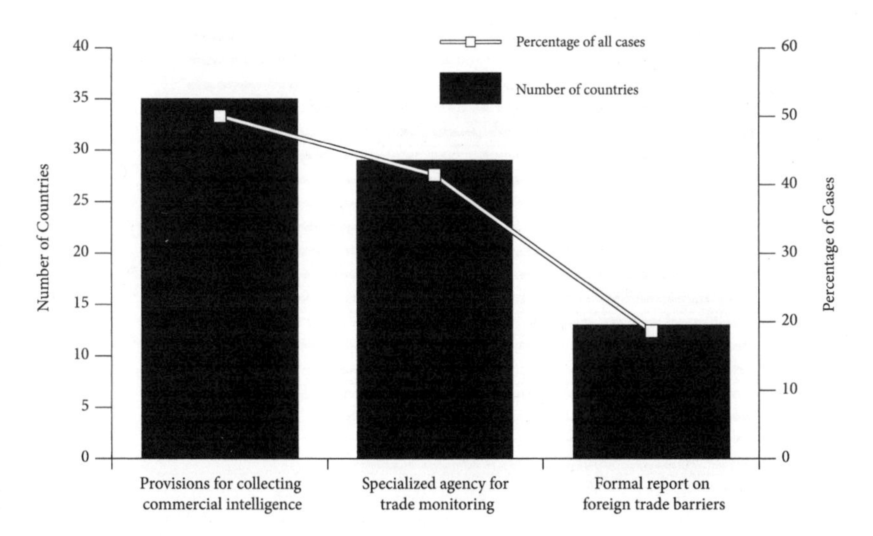

Fig. 32.1. Few WTO members have the capacity to monitor foreign trade barriers (2008). (Source: Ghosh, "See No Evil, Speak No Evil? The WTO, the Trade Policy Review Mechanism, and Developing Countries," D.Phil. Thesis, Oxford University, 2008)

in agriculture and India for anti-dumping measures). Wider use of trade measures would require a requisite increase in capacity for developing countries in general.

2. Strengthen WTO Monitoring of Protectionist Measures

A more efficient alternative to country-based monitoring is institutional monitoring by the WTO. The WTO's own Trade Policy Review Mechanism (TPRM) periodically reviews member states, based on WTO reports, government reports, and review meetings in which all members can participate. Although reviews are more frequent for the largest trading powers, even those only occur in two-year intervals. More significantly, thanks to resource limitations and a growing membership, the WTO has never managed to conduct the requisite number of reviews as required each year (Figure 32.2). Further, in only half the cases where developing

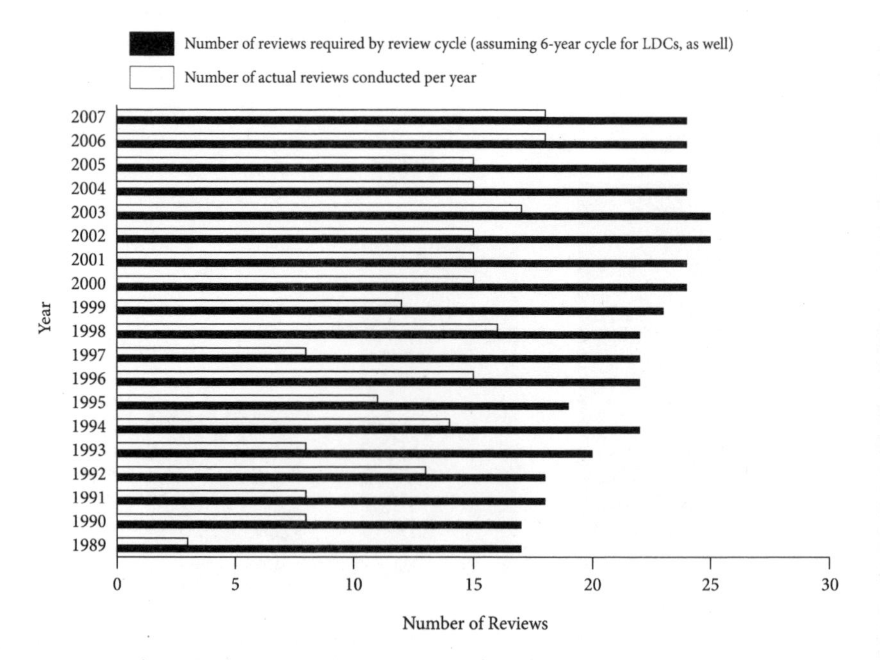

Fig. 32.2. The WTO's capacity to monitor trade policies is constrained. (Source: Ghosh, "See No Evil, Speak No Evil? The WTO, the Trade Policy Review Mechanism, and Developing Countries," D.Phil. Thesis, Oxford University, 2008)

countries formally challenged trade measures did the reports warn about the contentious measures in advance of the disputes. With this record, it is obvious that the monitoring of climate-related trade measures cannot be accomplished with existing resources or with the existing mandate in the WTO.

The WTO also has a system of notifications, whereby countries submit information every time new trade measures are introduced. But even rich countries often fail to submit notifications on time. Developing countries fear that gaps in notifications are deliberate strategies to withhold information.

More credible monitoring of climate-related measures would need, first and foremost, an increase in the resources allocated to the WTO. Increased resources would allow for more frequent monitoring by the TBT Committee and the Committee on Trade and Environment and more comprehensive reports under the TPRM. A second necessary reform would be to strengthen the notifications process by requiring countries to notify the WTO of climate-related measures prior to implementing them. This procedure has been adopted in new monitoring mechanisms within the WTO dealing with sanitary and phytosanitary standards (SPS) and regional trade agreements. A third requirement would be to ask countries to explain the rationale behind planned measures (again adopted for SPS monitoring). This would increase transparency, limit the cost to developing nations of challenging potentially unfair trade measures, and facilitate the ability of the wider WTO membership to apply pressure against contentious measures.

3. Define Categories for Environmental Goods and Services Clearly

To liberalize trade in EGS, environmental goods and services would need to be clearly defined. The WTO uses a six-digit level of product classification, which makes it difficult to distinguish between environmental goods and other products. It is also difficult to determine which products to liberalize when the product has multiple uses. Developing countries are unwilling to open up entire product categories to import competition. Similarly, trade measures to counter leakage would have to be targeted precisely at those products whose production methods are proven to adopt lower environmental standards. Poorly targeted measures would otherwise face charges of trade discrimination.

4. Overcome Measurement Challenges of Embodied Carbon across the Supply Chain

Another type of measurement difficulty arises from notions of embodied carbon, i.e., the amount of CO_2 emitted during each stage of a product's manufacturing and distribution to consumers. No standard methodology for this measurement has been adopted. Top-down analysis is difficult because sectoral averages could differ from the specific carbon-intensity of individual products. On the other hand, the level of detail required in bottom-up process examinations would impose capacity burdens on developing countries. Thus, even if methodologies were agreed upon, the capacity question would still need attention.

5. Build Developing Countries' Capacity to Monitor Emissions

The final issue relates more directly to the capacity of developing countries to measure emissions. Non–Annex I (NAI) parties submit inventories as part of their national communications, which do not include time series data and cover only CO_2, methane, and nitrous oxide. To date, although 134 NAI parties have submitted their first communications, even some of the largest developing country emitters have not submitted further reports (Figure 32.3). This is partly a strategic move to withhold information until a climate deal is agreed. But for many other developing countries, the self-reporting structure is under strain.

Building capacity to monitor emissions is not going to be easy. The United Nations Framework Convention on Climate Change's (UNFCCC) Consultative Group of Experts, which provided technical support to developing countries, allocated only USD 100,000 per country to monitor emissions. Its mandate expired in 2007 and was renewed only in June 2009. Although new centralized satellite technologies could measure emissions anywhere in the world, there is still a case for capacity building within individual countries. The climate regime is complex, and parties' willingness to participate would, in part, depend on their ability to monitor and verify data on their own without having to depend solely on data generated by rich countries or international organizations.

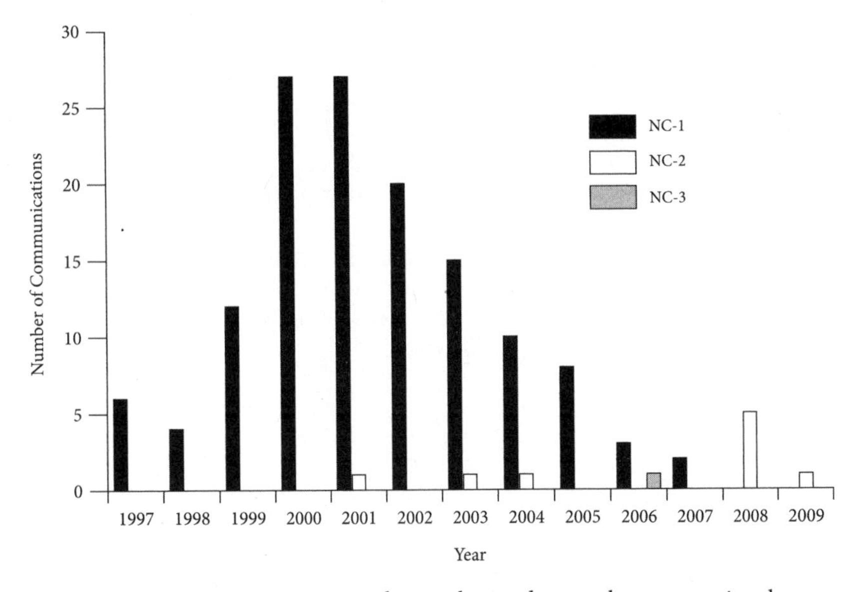

Fig. 32.3. Few Non–Annex I parties have submitted more than one national communication. (Source: Ghosh, "See No Evil, Speak No Evil? The WTO, the Trade Policy Review Mechanism, and Developing Countries," D.Phil. Thesis, Oxford University, 2008)

Conclusion

WTO rules and institutions are likely to become involved with climate rules in numerous ways. Developing countries have many concerns which, if not properly addressed, may limit the effectiveness or fairness of any global climate change agreement and of the WTO. It will be impossible to address many of these concerns unless transparent, effective, and fair monitoring systems are put into place. Otherwise, so-called efficient outcomes in climate negotiations might stumble during the implementation, monitoring, and enforcement stages.

FURTHER READING

For an overarching paper on links between environmental and trade policy, see Jeffrey Frankel, *Global Environmental Policy and Global Trade Policy* (Harvard Project on International Climate Agreements, October 2008).

A recent proposal suggesting positive linkages between standards and access to trade markets: Christian Barry and Sanjay Reddy, *International Trade and Labor Standards: A Proposal for Linkage* (New York: Columbia University Press, 2008).

For an in-depth analysis of the workings of the TPRM, see Arunabha Ghosh, "Information gaps, information systems, and the WTO's Trade Policy Review Mechanism," *Global Economic Governance Working Paper 2008/40*, (Oxford, May 2008).

For a review of the measurement, reporting, and verification arrangements in the climate regime, see Clare Breidenich and Daniel M. Bodansky, *Measurement, Reporting and Verification under the Bali Action Plan: Issues and Options* (Pew Center on Global Climate Change, April 2009).

NOTES

1. Veena Jha, "Environmental Priorities and Trade Policy for Environmental Goods: A Reality Check," ICTSD Issue Paper 7, September 2008.
2. Arunabha Ghosh, "See No Evil, Speak No Evil? The WTO, the Trade Policy Review Mechanism, and Developing Countries," Oxford University, D.Phil. Thesis, 2008, pp. 257–262.

Chapter 33

Carbon Footprint Labeling in Climate Finance

Governance and Trade Challenges of Calculating Products' Carbon Content

Sandra G. Mayson

Scholar, Institute for International Law and Justice, NYU School of Law

Key Points

- Carbon footprint labeling (CFL) attempts to quantify the GHG emissions attributable to a product throughout its life cycle, from the harvesting of raw materials through product disposal.
- CFL could impose an increased regulatory burden on small producers and a relatively greater abatement burden on developing countries.
- A number of CFL standards have already emerged, backed by governments, NGOs, industry groups, and the ISO. Divergent choices in calculation methodologies (what emissions a CFL covers and how they are measured) have contributed to this multiplicity.
- Governments seeking to ensure that mandatory national CFL programs are WTO-compliant should adopt a sound international CFL standard, created with wide national and stakeholder participation and sufficiently flexible to accommodate individualized producer data.

Yesterday, it was trans-fat; today, carbon footprint labels are proliferating on grocery store shelves. Carbon footprint labels purport to quantify the embodied carbon of a given product: the total quantity of carbon dioxide and (in some cases) other greenhouse gases (GHG) for which a single product—a pear, a cell phone, a t-shirt—is responsible over the course of its life cycle, from creation through use and disposal. Carbon footprint labeling (CFL) is a new phenomenon but has already staked a place in the climate regulatory landscape. Viewed most optimistically, CFL harnesses consumer demand for low-carbon products to encourage emissions reductions down supply chains. Critics, however, see CFL as a form of disguised protectionism, devised by industry or well-meaning nongovernmental organizations (NGOs) and promoted by governments in the developed North to counter the comparative advantage of producers in the global South subject to less stringent emissions controls.

CFL may serve as a valuable informational tool to promote awareness about products' emissions costs. Early evidence suggests that product footprint labeling helps firms to identify CO_2 emissions hotspots along supply chains. CFL is also intended, however, to attach a cost to greenhouse gas emissions. If consumers respond to carbon labels in purchasing decisions, CFL should result in a loss of market share for high-emissions goods and services, and create market access (or advantage) for goods and services with low carbon content. By one view, this is a form of protectionism—at least if CFL is mandated by governments. The difficulty of quantifying carbon content compounds the risk that CFL might distort markets, or strain other climate law regimes by creating separate incentives for emissions reductions. Critics also fear that carbon labels will distract from other externalities of production and consumption.

Given their regulatory and distributional implications, the development of CFL standards deserves close attention. Who decides how to calculate embodied carbon? NGOs and industry have taken the lead to date. Their labeling standards could, through market impact down supply chains, have significant effects on climate finance—yet they operate largely independently of international climate agreements and official state measures. This situation raises important questions about the governance and accountability of CFL standardization processes. It also makes the analysis of CFL's legality under the World Trade Organization's (WTO) trade regulatory disciplines complex, since it depends, in part, on whether labeling programs are mandated or promoted by governments or established solely by non-state actors.

The Rise of Carbon Footprint Labeling

Developing a carbon label is no simple task. Labels take different forms. Comparative labels simply present information about a product's embodied emissions, like a food nutrition label. Endorsement labels signify that a product's embodied emissions fall below a given threshold. Organizations that issue labels may require emissions reductions or third-party verification as a condition of the label's use.

Calculating the emissions for which a single product is responsible requires choices about what to measure (the "system boundary") and how. Will the calculation include emissions from machinery used to harvest raw materials? From factories that produce the machinery? From land use change? Worker transport? What level of data specificity will be required? A Life Cycle Analysis approach requires individual source data, while environmental input-output (EIO) analysis uses sector-level national averages. Label designers must also decide how to account for the fact that the emissions might vary according to the user's choices (e.g., to recycle or not) and context (e.g., local energy grid).

Critics contend that these and other conundrums make it impossible to accurately quantify a product's carbon content. The variables are simply too uncertain, and the methodological choices too arbitrary. A myopic fixation on carbon footprints, moreover, may distract from other environmental and social costs of production. Others argue that complex, costly labeling standards impose a disproportionate burden on small producers and circumvent the principle of common-but-differentiated responsibilities, since—international treaty agreements notwithstanding—producers in developing economies must either monitor and reduce emissions or lose market share.

Despite such concerns, carbon labels are multiplying. While other environmental and social labeling programs took decades to evolve, CFL has become an international phenomenon in the space of a few years. The pioneer initiatives have been hybrid private-public projects, though some NGO and industry efforts are progressing with no state involvement at all.

The most advanced CFL regime is Publicly Available Standard (PAS) 2050, designed by the British Standards Institute in collaboration with the British government's Department for Environment, Food, and Rural Affairs (DEFRA) and the Carbon Trust, a government-funded NGO. Two other hybrid CFL initiatives are vying for international status: One launched by the Greenhouse Gas Protocol (a partnership between the

NGO World Resources Institute and the industry collective World Business Council for Sustainable Development), which developed a successful set of corporate accounting standards for GHG emissions; the other by the International Organization for Standardization (ISO) (which essentially adopted the GHG Protocol's corporate emissions accounting standard in 2006).

At the national level, ten German corporations have joined forces with the World Wide Fund for Nature (WWF) and two academic institutes to develop a labeling standard. The US-based NGO Carbon Fund and Canadian NGO CarbonCounted are certifying low-carbon products. Swedish organic standards association Krav has a label underway. Industry-sponsored labels include those developed by French supermarket chains Casino and E. Leclerc and Switzerland's Migros.

Governments are increasingly promoting CFL. Japan and South Korea have both announced plans for government-run labeling regimes. The British government has been integrally involved in the development of PAS 2050; the Greenhouse Gas Protocol's Steering Committee includes government agencies from a handful of countries; and the ISO is composed of national delegations. The European Parliament has called for the development of data to enable GHG footprint labeling (including on imports) and is developing a Carbon Footprint Measurement Toolkit. If the US Congress passes legislation requiring border tax adjustments based on products' embodied carbon, it will have to address carbon footprinting as well. The California legislature, finally, is considering the proposed Carbon Labeling Act, which would require the state to create and implement a (voluntary) carbon labeling program.

Harmonization of CFL Standards?

The short history of CFL illustrates conflicting trends: diversification among labels and a drive towards uniformity. Almost every institution that has launched its own footprinting initiative has simultaneously pled for harmonization. There is no strong evidence of convergence thus far, but many CFL standards overlap, and market and political pressures may propel a few—or even a single standard—to preeminence. The emergence of a dominant CFL standard could lower implementation costs and mitigate CFL's potentially disproportionate burden on small producers and developing economies. The precise terms of any such standard, however, would have varying competitiveness implications for different countries and firms.

While PAS 2050 may provide a basis for a universal standard, the GHG Protocol and ISO appear most likely to achieve it. The GHG Protocol's explicit objective is to create a harmonized international standard, which it hopes the ISO will adopt. Given the success of the GHG Protocol and ISO accounting standards, the ISO's international profile, and the GHG Protocol's careful multi-stakeholder process, a GHG Protocol/ISO product footprinting standard could well dominate the field.

CFL and the WTO

The WTO TBT Agreement requires that technical standards, which would include carbon footprint labeling standards, that are adopted or mandated by governments must conform to the procedural and substantive requirements norms for standard setting provided in the TBT Annex 3 Code of Good Practice. Technical regulations are required to be non-discriminatory and "not more trade-restrictive than necessary to fulfill a legitimate objective." In the case of domestic or regional voluntary standards adopted by non-governmental bodies, WTO members are obliged to take "such reasonable measures as are available to them" to ensure compliance with Annex 3 norms; this obligation does not extend to international standards. It is unclear what degree of government involvement or endorsement might be sufficient to make the TBT disciplines directly applicable to standard setting by a private body. Would a private program be subject to challenge if a government sets mandatory criteria for, or regulates access to, a carbon label? Would the UK's sponsorship of PAS 2050 (via the Carbon Trust) suffice? WTO law and jurisprudence offer scant guidance on these questions.

CFL standards may also engage TBT provisions establishing that when a WTO member country bases a technical regulation on "relevant international standards" set by a "recognized body," it enjoys a presumption of legality. The TBT does not define "relevant international standard" or "recognized" standard-setting body. Nor does it address a situation of competing standards. A WTO member that adopted a private CFL standard could well seek to invoke the presumption, requiring a WTO dispute settlement panel and the Appellate Body to clarify these issues. Given the economic and environmental stakes of CFL standard-setting, it would be appropriate for a WTO tribunal tasked with deciding whether or not to extend a presumption to a private CFL standard to determine whether the standard-setting process that produced it is transparent, whether it allows for meaningful participation by affected interests, and whether the standard-setting body justifies decisions by public reasons and evidence. Other TBT provisions that might be relevant to the legitimacy of CFL regimes include the Agreement's code on conformity assessment procedures; its exhortation to allow market access to goods that comply with exporting countries' regulatory standards; and the obligation of developed countries to assist developing country producers to comply with labeling requirements.

Alternatively, a government might use CFL standards or methodologies to exclude certain products with high carbon footprints, or impose a tax on products with heavier footprints. Such a regulation would not only be subject to the TBT disciplines but also potentially be subject to challenge as discriminatory under the GATT. The central issue would be whether products with heavier footprints are "like" similar products with lighter footprints. If so, they must be treated the same unless the government imposing the label can justify the disparate treatment. The question is whether the methods by which a product is produced, consumed, and disposed of—as opposed to the physical characteristics of the product itself—are relevant in determining likeness. Given that consumers may differentiate between products with varying levels of embodied emissions, there is a reasonable argument that heavy-footprint products are not "like" light-footprint products under the GATT. Even if the products at issue were deemed like, government measures treating them differently might still pass muster on a showing that the measures are necessary to protect human, animal, or plant life or relate to the conservation of exhaustible natural resources. Such a justification would require, among other criteria, that a labeling regime be procedurally fair and flexible enough to accommodate divergent practices among producers. An Environmental Input-Output (EIO) methodology based on national sectoral emissions averages might fail.

Conclusion

Whether carbon labels come to function as de facto conditionalities on investment or just help to shape a low-carbon culture, it seems clear that they will remain one element of the emerging matrix of climate finance. The development of carbon labels by hybrid public-private bodies presents a challenge for accountability in international governance. Given the special trade stature of international standards, inclusiveness in international CFL initiatives is paramount. Broad participation might help also mitigate CFL's distributional impact. These dictates of good governance also align with the objective of developing WTO-compliant national labeling regimes. The closer a labeling requirement is to a widely endorsed international standard, and the more adaptable to individual producer data, the more likely the labeling program will be to pass muster under WTO law. That general principle in mind, carbon footprint labeling is a new phenomenon. Climate professionals will have to continue to assess the effect of carbon labels on other emissions reductions regimes, as well as their trade law status, as labeling regimes evolve.

FURTHER READING

A. Appleton, "Private Climate Change Standards and Labelling Schemes under the WTO Agreement on Technical Barriers to Trade," *World Trade Forum* (2007).

Paul Brenton et al., "Carbon Labeling and Low-Income Country Exports: A Review of the Development Issues," 27 *Development Policy Review* 243 (2009).

Jiang Kejun et al., *Embedded Carbon in Traded Goods*, ICTSD Background Paper (2008).

Part VI

Taxation of Carbon Markets

Chapter 34

Fiscal Considerations in Curbing Climate Change

Lily Batchelder
Professor, NYU School of Law

Key Points

- The choice between cap-and-trade and a carbon tax should mostly be made on political grounds, focusing on whether the targeted price change or emissions level is clearer, the likelihood of accurate distributional offsets, budgetary conventions, agency competence, and the salience of the cost imposed.
- Climate regimes are highly regressive, disproportionately burdening the least well-off. Offsetting these distributional impacts is desirable purely from an efficiency perspective, and also because such regressivity undercuts one of the fundamental goals of curbing climate change.
- Carbon taxes are likely to raise more revenue than cap-and-trade schemes to mitigate distributional effects because of the political tendency to allocate many permits for free. Free permits run the risk of benefiting the owners of politically savvy emitters, rather than those who are actually burdened. Funds from both schemes, however, may fail to reach those most affected, including the elderly, disabled, working poor, and unemployed.
- Domestically, distributional offsets are more likely to be sufficiently large and well-targeted if structured as a universal contributory scheme, with all carbon revenues transparently used to fund direct rebates for all. Internationally, reasonable approaches include

gradual extension of permitting or tax regimes to less developed countries coupled with international carbon offsets, or excess permit allocations based on an objective measure of fiscal capacity.

Introduction

Climate change abounds with fiscal issues. At a macro level, the debate between a carbon tax, cap-and-trade system, and command-and-control regulation is about the extent to which the tax system is the best vehicle to address climate policy objectives. At a micro level, energy-related fiscal incentives and the tax treatment of carbon taxes, carbon permits, and climate markets can have important implications for a regime's effectiveness. The question of how to address the distributional impacts of carbon mitigation, both domestically and internationally, is also a fiscal issue.

This chapter provides a brief summary of the fiscal, administrative, and political considerations relevant in designing a climate mitigation regime. It then focuses on the importance of distributional offsets, and the challenges in implementing them. Other fiscal issues, including the nuts and bolts of taking carbon permits and carbon markets, are addressed by Kane (chap. 35) and Margalioth (chap. 36).

Fiscal Issues in Climate Regulatory Choices

While climate change policy can be, and is, implemented through a variety of mechanisms, including fiscal subsidies and command-and-control regulation, the current debate rightfully focuses on carbon taxes and cap-and-trade systems. Because the two can theoretically be structured to be economically equivalent, the decisive issues are political—how each will realistically be enacted and implemented.

Keohane (chap. 5) outlines two critical considerations. Because the damages from climate change appear to rise sharply above some emissions level, cap-and-trade regimes can minimize externalities with fewer adjustment costs. Allowing permit banking can address permit price volatility under a cap-and-trade scheme. In addition, the fact that a carbon tax is denominated as a "tax" may generate more political opposition and thus limit its scale. Nevertheless, three additional fiscal issues, described below, are usually overlooked and highlight that there may be no one right answer. The best choice between carbon taxes and cap-and-trade will vary by country, and may be a hybrid of the two.

Domestic Budgetary Conventions

How a climate mitigation regime will be treated under a country's budgetary conventions and procedures may be important when selecting a regime. The EU, for example, requires a unanimous vote for tax legislation, but only a majority vote for other bills. As a result, it has adopted a cap-and-trade regime, which policymakers were careful to ensure was not categorized as a tax. In other countries, however, enacting tax legislation is typically easier. For example, the US periodically requires fully paying for the cost of any legislation with revenue raisers. Costs and revenue raisers are calculated over a five- or ten-year budget window. These rules tend to make it easier to pass tax legislation because the tax committees control which revenue raisers are passed. They also artificially reduce the budgetary cost of legislation that raises revenue in the short term while deferring costs to the long term. Cap-and-trade regimes are more likely to grandfather existing emitters in the short term, which artificially inflates their budgetary cost, and are not treated as taxes. Thus, they may be more difficult to enact in a US-style budgetary environment.

Domestic Administering Agency

States must also consider what agency can administer the regime most efficiently. Revenue agencies usually take the lead on carbon taxes, while environmental agencies take the lead on permitting regimes. Revenue agencies have the advantage of extensive experience in auditing and collection, and typically administer energy-related taxes and subsidies already. But their primary focus is on measuring income, not emissions. An environmental agency, by contrast, may focus more narrowly on this dimension and obtain higher compliance rates. However, these differences are probably overstated. Countries are increasingly giving substantial responsibility to other agencies when administering tax programs. Likewise, permitting agencies can verify carbon use more effectively if they partially rely on information from revenue agencies on firms' income and deductions.

Offsets to Mitigate Distributional Inequity

Because the impact of any climate regulatory regime is likely to be strongly regressive, a final important issue is what distributional offsets are likely to accompany each approach. As explained below, such offsets are desirable purely on efficiency grounds. They are also necessary from an equity perspective, even if one disregards historical contributions to climate change and claims that the current global economic distribution is unjust. In addition, they are important practically. While low-income individuals and countries typically have less political influence, they may nevertheless block enactment of a climate regime that disproportionately burdens them.

Carbon taxes are likely to raise more revenue that can be used for such offsets. Allocating free permits under a cap-and-trade system is another way to limit the distributional impacts. But it is less well targeted because much of the value accrues to investors in recipient firms, rather than the consumers burdened. Free permits can also result in inequities and inefficiencies if some industries and countries obtain them for emission reduction efforts that they would have undertaken absent the regime. Despite the greater revenue generated by a carbon tax, however, it may be difficult politically to direct such revenue to those most affected, as discussed next.

Addressing the Regressivity of Climate Mitigation

Offsetting the distributional effects of a climate regime is critical for two reasons. First, these effects undercut one of the fundamental rationales for curbing climate change—avoiding the increased rate of poverty and preventable deaths that scientists project if we continue on our current emissions path. In 2000, the World Health Organization (WHO) estimated that climate change already cost about 5.5 million disability-adjusted years of life annually. Stern and others project that further climate change will result in a roughly 11% reduction in global GDP, and large increases in infectious diseases and malnutrition. The total disease burden will be borne largely by children in developing countries. This creates a strong imperative to act now. As John Roemer argues, there is little reason to weight the utility of current generations more heavily than future generations.

If the distributional effects of climate mitigation are not offset, however, the regime may increase poverty and preventable deaths on net. Most economists agree that the burden of any climate regime will be borne largely by low-income individuals and, if it is multilateral, individuals in developing countries. About half of the world's population lives on less than $2 per day. Largely as a result, an immense number of people already die of preventable deaths each year. For example, the WHO estimated that malnutrition cost roughly 138 million disability-adjusted years of life in 2000, and unsafe water, sanitation, and hygiene cost about 54 million. The vast majority of these deaths and disabilities were children in developing countries.

Because climate regimes tend to be regressive on a national and global level, they will increase short-term global poverty absent large and well-targeted offsets. At the extreme, this possibility implies that we should do less to mitigate climate change if distributional offsets are not enacted at the same time. Put differently, if there are no distributional offsets, we would be addressing the catastrophic costs that climate change imposes on future generations by imposing greater catastrophic costs on the most vulnerable individuals in the present.

Second, even if these considerations are disregarded, the distributional effects of climate regimes should be fully offset purely on efficiency grounds. Analyzing the efficiency of a policy change requires holding distributional preferences constant. All societies have distributional preferences, and redistribution entails efficiency costs. If distributional preferences were not held constant, one could argue that all regressive policy changes (say, a subsidy for private yachts) were efficient, even if the only efficiency benefits stem from an assumption that society's distributional preferences have changed.

In the climate change context, there are two options for holding the level of redistribution constant. One is to use all of the revenues potentially generated by the regime to fund transfers offsetting its regressive effects. Another is to use these rents in other ways (say, to buy off interest groups), and raise existing taxes to fund the even larger transfers necessary to hold the level of redistribution constant. The latter approach entails efficiency costs because it increases the distortions the tax system already imposes on the choice between labor and leisure. Thus, the only way to avoid efficiency costs is to use climate revenues directly to offset the scheme's distributional effects. This is true even under the assumption that the current global distribution is fair.

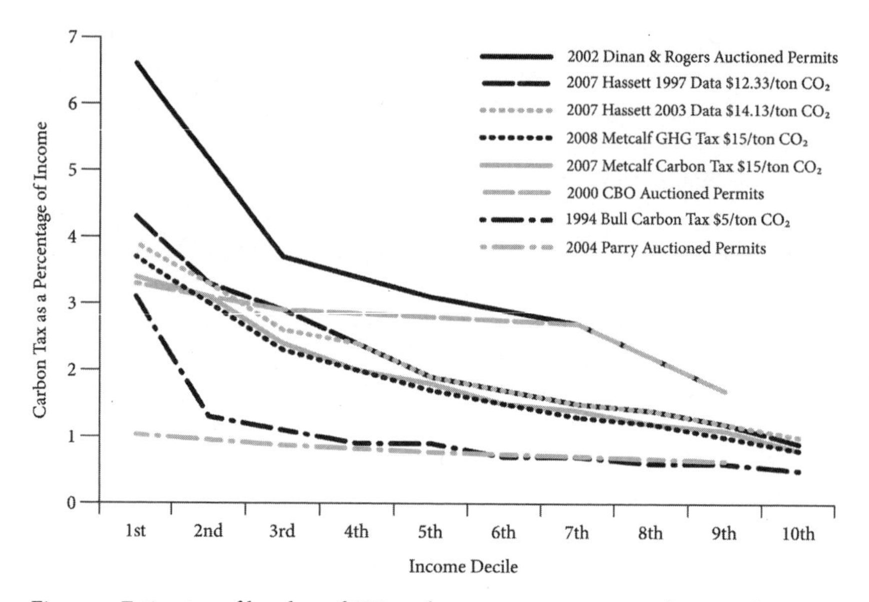

Fig. 34.1. Estimates of burden of U.S. carbon tax as percentage of income by decile. (Source: Tracey M. Roberts, *Mitigating the Distributional Impacts of Climate Change Policy* (mimeo, May 1, 2009))

As illustrated by Figure 34.1, climate regimes are indeed highly regressive, even in a purely domestic setting. Lower-income households bear a larger burden as a percentage of their income because they tend to spend a larger share of their income on carbon-intensive products. This is also true in other nations and across countries.

a. Challenges in Enacting Distributional Offsets

A number of political dynamics may limit the ability to offset these distributional effects of climate regimes. First, experience suggests that, at least initially, cap-and-trade systems tend to allocate most permits to existing emitters for free. Theoretically, this could result in permitting regimes addressing distributional effects more reliably. After all, direct transfers and foreign aid are often stigmatized as welfare. But if free permits are allocated disproportionately to some firms, such as those that are old, large, or politically savvy, they will generate sharp differences in the costs of the regime for competing companies. Firms receiving free permits may be able to raise their prices in the short term by the same amount as their competitors. Then much of the benefit of these free permits would accrue to the owners of such firms in the short term. Those burdened by the regime—ordinary consumers—would obtain relatively few benefits.

In addition, any funds that are raised by carbon taxes or auctioning permits may fail to reach the groups most affected by climate change policies. These include the elderly, disabled, working poor, and unemployed. While some developed countries may be willing to provide direct transfers to such households, others, like the US, may resist doing so. There is traditionally strong opposition in the US to transfers that are not conditional on work. Offsets delivered through the tax system may fare better, but they also present political challenges. For example, most income tax subsidies in the US take the form of deductions, exclusions, and nonrefundable credits. These subsidies provide few benefits to households in lower tax brackets, and none to those with no income tax liability—roughly 40% of US households. The only tax benefits that can reach such households are refundable tax credits, but these are also difficult to enact politically.

Offsetting the distributional impact of a climate regime internationally will be even harder. There is strong opposition to increasing foreign aid in many developed countries. For example, according to the Congressional Research Service and OMB, the US spends about 1.2% of its discretionary budget on foreign aid aimed at poverty reduction, much less than the roughly 28 percent spent on domestic income security programs. Voters may be even more resistant to international offsets if they involve cutting back on domestic distributional offsets that they have come to view as an entitlement.

b. Steps to Enhance the Efficacy of Distributional Offsets

Despite these challenges, past experience does imply at least two ways to improve the sufficiency and accuracy of distributional offsets that policymakers should consider.

First, experience with domestic programs like government pensions (e.g., Social Security in the US) suggests that earmarking a discrete revenue source for transfers structured as a universal contributory scheme can protect a program over time. Under a universal contributory scheme, all receive transfers linked to contributions. Because all benefit, and benefits are linked to burdens, they tend to garner more widespread political support.

This experience suggests that the revenue raised by any climate regime should be dedicated exclusively to distributional offsets. The funds raised should be rebated to all households, not just those bearing the largest burdens, perhaps as a flat dollar amount per person. Prices on consumer goods should also separately state the embedded cost of carbon taxes or permits so that it is clear that all are contributing.

Second, opaque redistributive transfers appear to garner more political support than transparent ones. This would imply allocating free permits in the domestic context, but, as explained above, they are likely to be poorly targeted. Instead, issuing domestic rebates through the tax system as refundable tax credits is probably the better approach.

Internationally, excess permit allocations may be the only option that is politically viable, given the aversion in developed countries to spending on foreign aid. However, doing so raises targeting issues similar to those in the domestic context. If excess permits are allocated disproportionately to some developing countries, or some firms operating there, the principal beneficiaries in the short term may be the investors in such countries and firms. The consumers who are burdened would benefit relatively little. As a result, policymakers should consider using an objective measure of fiscal capacity, such as per capita income, to allocate excess permits. Allowing low-income countries not to participate in tax or permitting regimes may also be effective from a distributional perspective. Non-participation could undercut environmental goals given the large share of emissions from the developing world. But if firms could purchase carbon offsets in such countries in lieu of purchasing permits or paying taxes domestically, there would still be incentives to reduce emissions in non-participating countries. As discussed elsewhere in this volume, such international carbon offsets are prone to gaming, but new forms may be more effective.

FURTHER READING

Joseph E. Aldy, Alan J. Krupnick, Richard G. Newell, Ian W. H. Parry, and William A. Pizer, *Designing Climate Mitigation Policy* (RFF Discussion Paper 8-16, May 2009).

Terry M. Dinan, and Diane Lim Rogers, "Distributional Effects of Carbon Allowance Trading: How Government Decisions Determine Winners and Losers," 55 *National Tax Journal* 199 (2002).

Robert Greenstein, Sharon Parrott, and Arloc Sherman, *Designing Climate-Change*

Legislation That Shields Low-Income Households from Increasing Poverty and Hardship (Center on Budget and Policy Priorities, May 9, 2008).

A. Haines, R. S. Kovats, D. Campbell-Lendrum, and C. Corvalan, "Climate Change and Human Health: Impacts, Vulnerability and Mitigation," 367 *Lancet* 2101 (2006).

Joint Committee on Taxation, *Climate Change Legislation: Tax Considerations* (JCX29-09, June 12, 2009).

John E. Roemer, *The Ethics of Distribution in a Warming Planet* (Cowles Foundation Discussion Paper No. 1693, April 2009).

Chapter 35

Tax and Efficiency under Global Cap-and-Trade

Mitchell A. Kane
Professor, NYU School of Law

Key Points

- Two approaches to the taxation of carbon markets and abatement opportunities can be taken to avoid distorting the market and its participants' behavior and thereby to preserve the efficiency of trading-based climate regulatory systems: inter-firm tax neutrality and intra-firm tax neutrality.
- Inter-firm tax neutrality requires that all abatement costs receive the same tax treatment and that all permits receive the same tax treatment, regardless of the firm which undertakes the abatement or acquires permits. In the context of international emissions trading, this approach requires harmonization of the respective domestic tax rates for permits and abatement.
- Intra-firm tax neutrality requires that each firm face the same tax treatment of actual abatement and permits on the margin. In the context of international emissions trading, this approach requires each country to achieve this matching, but does not require harmonization of tax rates. It does not require international harmonization of tax systems.
- In the real world, intra-firm tax neutrality is the preferred policy approach due to the lesser degree of required coordination among national tax systems. The key challenge in implementing intra-firm tax neutrality will be to match tax treatment of permits and abatement. Because permits are likely to receive the same tax treatment

> for all holders, this means the efficient tax policy will require removing national-level tax differences among different methods of abatement (except where they are justified by non-climate externalities) or making them ineffective at the margin. Coordination of this particular tax policy goal would be best achieved under the aegis of international climate agreements rather than through tax treaties.

A cap-and-trade regime relies on the price of permits to signal which abatement opportunities are cost-effective, in light of the overall cap. Just like any market where we use price signals to achieve allocative efficiency, taxation is a looming problem. To the extent that taxes distort prices, the market will not function optimally, impairing the efficiency of the regulatory system. The very fact that one requires a market to achieve efficient abatement in the first place only arises because there are firm-specific low-cost abatement opportunities. Such firm-specific opportunities can take one of two forms. First, some firms may have low-cost abatement opportunities due to the ownership of some type of proprietary technology that allows production with fewer emissions than competitors. Second, some firms may have low-cost abatement opportunities because they happen to operate in jurisdictions where there are relatively low-cost abatement opportunities. Taxation presents the same type of potential problem in each of these cases: abatement opportunities that should be favored on a pre-tax basis become relatively expensive on an after-tax basis due to differential tax treatment of firms operating in the market, either due to their mode of production/abatement or their territorial location of operations. (Some tax preferences might be independently justified by non-climate externalities, such as national security, and would accordingly not distort the market; these preferences are not the subject of the analysis which follows.) In a first best world there are two ways to structure tax systems in order to preserve efficient allocation of abatement. Each approach involves concepts of tax neutrality, but they operate at different levels. Thus, we can distinguish between inter-firm tax neutrality and intra-firm tax neutrality.

Inter-firm Tax Neutrality

Inter-firm tax neutrality is the more intuitive of the two types of tax neutrality, albeit the form that is much more difficult to achieve in a multi-jurisdictional cap-and-trade system. The goal is to remove tax distortions that operate to shift abatement away from firms which have low-cost abatement opportunities and toward firms with high-cost abatement. For example, suppose that Firm A can abate a ton of carbon emissions at a cost of USD 20 and Firm B can abate a ton of carbon emissions at a cost of USD 15. If we imagine that Firm A faces a 50% marginal tax rate and Firm B faces a 10% marginal tax rate, then the after-tax cost of abatement (which should give rise to a deductible expense under standard income tax principles), will be USD 10 and USD 13.50, respectively. All else equal, Firm A will inefficiently abate on the margin instead of Firm B. To remove the distortion it would be necessary to ensure that Firm A and Firm B face the same tax rate on their abatement expenses. By itself, however, this condition would not be sufficient because there is another aspect of the market that may give rise to tax differentials. Specifically, firms may face differential tax treatment of permits (e.g., the acquisition cost of permits might be deductible by each of two firms but at different rates). If we conceive of acquiring and holding permits as the functional equivalent of *not* abating, then what we require under inter-firm tax neutrality is that all firms face the same tax treatment with respect to (i) actual costs of abatement and (ii) actual costs of *not* abating (i.e., acquiring or retaining permits and using them to cover emissions). Note, however, that inter-firm tax neutrality does *not* require that we tax actual abatement and permits the same as each other. If they are taxed differentially, then in a liquid market we should observe equilibrium price effects on the price of permits (which will capitalize the tax benefit or detriment relative to the tax treatment of actual abatement), but there would be no reason for abatement to shift inefficiently across firms, as no firm has an advantage relative to any other firm with respect to either abating or not abating. The chief problem in achieving inter-firm tax neutrality is that it would require an unprecedented degree of harmonization of tax rates and bases across the world.

Intra-firm Tax Neutrality

A different type of tax neutrality which could be substantially more feasible to implement might be termed intra-firm tax neutrality. The intuition here is that if every firm in the market is made tax indifferent on the margin between abating and not abating (i.e., acquiring and holding permits), then the market taken as a whole should be efficient. The condition required to implement this form of neutrality is that any given firm face the same tax treatment of actual abatement costs and the permits that operate as substitutes for that abatement. This condition does not require that a given firm face the same tax rate on all possible methods of abatement and permits that it might acquire. The point rather is that when a source faces the choice between particular methods of abatement versus holding an additional permit on the margin, then the tax treatment of such abatement and of such permit should be the same. This is crucial because the condition can be satisfied without harmonization of tax rates across countries. Thus if Firm A operates in Jurisdiction 1 and Jurisdiction 2, intra-firm tax neutrality does not require that it face the same tax rate on abatement and on permits in Jurisdiction 1 and Jurisdiction 2. Rather, all that is required is the same treatment of abatement and of permits within each jurisdiction, i.e., that Firm A face the same tax treatment on (i) permits held for surrender to Jurisdiction 1 and actual costs of abatement which reduce emissions in Jurisdiction 1 and (ii) permits held for surrender to Jurisdiction 2 and actual costs of abatement which reduce emissions in Jurisdiction 2, and so on.

The Pragmatic Policy Solution

Intra-firm tax neutrality is the superior tax policy solution for minimizing market distortions and regulatory inefficiency because it can be implemented without tax rate and base harmonization across countries, which would be impossible to achieve. The key problem in achieving intra-firm tax neutrality is that governments, responding to powerful political pressures, will inevitably give tax credits or other preferences for particular abatement technologies or activities. Permits are very likely to receive uniform tax treatment (e.g., a straight deduction at the taxpayer's marginal tax rate in the period that the permit is surrendered). But variations in the treatment of abatement costs mean that national tax systems will never successfully achieve complete matching of abatement and permit costs. Nonetheless, intra-firm tax neutrality requires only that firms face the same tax treatment for permits and actual abatement on the margin. Infra-marginal tax differentials do not matter. Thus, it is possible to achieve intra-firm tax neutrality in the presence of tax subsidies for particular abatement methods, so long as the subsidies are fully exhausted short of the margin at which firms choose between abatement and permits. For example, if a country gave tax credits for solar energy, intra-firm neutrality would be achieved so long as the program is designed in a way such that any firm that takes advantage of it exhausts its allotment of credits prior to the point at which it must decide between further abatement and holding permits.

In the context of domestic climate trading systems, a country can successfully achieve intra-firm neutrality without harmonization of tax systems across countries. In the case of trading systems operating among states, coordination among countries is needed. The goal is not to harmonize rates or bases but to agree that national tax preferences regarding abatement should be designed to operate only infra-marginally. Moreover, universal coordination is not necessary to attain benefits, which will arise as each additional country adopts the preferred policy. Because the ultimate objective is adoption by all countries of the same policy, coordination is more likely to be achieved under the aegis of a multilateral climate agreement, rather than through the fragmented processes of bilateral tax treaties. The climate framework agreement is also the preferable forum because the tax policy goal in question has important substantive implications for the efficient and equitable functioning of international emissions trading. If one or more countries fail to follow the intra-firm neutrality norm, for example, by maintaining tax preferences that are effective at the margin, then we will observe too much abatement in those countries as compared to the efficient outcome. Moreover, the effect will be to deflate worldwide permit prices because equilibrium marginal abatement costs will be depressed due to the tax preferences. Countries that are net permit exporters would thus bear a cost in terms of lower permit revenue. Thus, the coordination of tax policy with respect to trading systems has efficiency and distributional consequences that go to the core of climate policy and climate politics.

FURTHER READING

Thomas Eichner and Ruediger Pethig, *Efficient CO_2 Emissions Control with National Emissions Taxes and International Emissions Trading*, CESIfo Working Paper No. 1967 (2007).

Carolyn Fischer, "Multinational Taxation and International Emissions Trading," *28 Resources and Energy Economics 139* (2006).

Ethan Yale, "Taxing Cap and Trade Environmental Regulation," *37 Journal of Legal Studies 535* (2008).

Chapter 36

Tax Consequences of Carbon Cap-and-Trade Schemes

Free Permits and Auctioned Permits

Yoram Margalioth

Visiting Professor, NYU School of Law;
Professor, Tel Aviv University School of Law

Key Points

- The tax treatment of cap-and-trade permits can distort permit markets and thereby undermine regulatory efficiency; tax rules should be designed and if necessary modified to avoid these problems.
- In terms of basic tax treatment, abatement and permit (upon surrender to the government) costs should be deductible from gross income. No depreciation deduction for permits should be allowed. Any gains made by selling permits should be taxable capital gains, unless the seller carries this out as a business, in which case these gains are ordinary income.
- The lock-in effect of imposing taxes only when an asset is sold is exacerbated when permits are allocated gratis, distorting permit prices. This effect can be reduced by auctioning permits (with the additional potential of using the proceeds to moderate regressivity) or by taxing the permits upon receipt.
- First-in-first-out and inventory accounting (using a mark-to-market basis) can help reduce additional lock-in distortions associated with fluctuating permit prices.
- Making the tax system symmetric with respect to permit gains and losses will reduce price volatility and resulting lock-in and other inefficiencies.

- Cap-and-trade increases the importance of transfer pricing rules to prevent market distortions arising through tax arbitrage strategies by multinational firms seeking to exploit differences across jurisdictions in the taxation of permits.

The cap-and-trade system creates a new asset—the permit. The tax treatment of permits can potentially distort the tradeoffs that sources make between abating or holding permits to cover their emissions, and thereby impair the efficiency of the regulatory system. This chapter first outlines the appropriate general income tax treatment of permits. It then addresses ways of dealing with the intensified lock-in effects and inefficiencies created by the current tax system's treatment of gratis permit allocations—by inventory management practices in the face of permit price fluctuations and by asymmetric tax treatment of permit gains and losses. It also addresses transfer pricing problems arising out of multinational firms' arbitrage among differences in the taxation of permits in the different jurisdictions in which they operate.

The Appropriate Basic Income Tax Treatment of Permits

Business expenses, the costs incurred by the taxpayer in the production of income, must be deductible if the income tax is to be imposed on income and not on sales, thereby becoming an excise tax on transactions. In the US, for example, section 162(a) of the Internal Revenue Code authorizes the deduction of "all ordinary and necessary expenses incurred during the taxable year in carrying on any trade or business." Abatement costs incurred in order to produce business income in compliance with the law clearly fall into this category and should be deducted from gross income. Instead of incurring abatement costs, the taxpayer can obtain, hold, and in due course surrender to the government a permit to cover its emissions at the end of the year in which they occurred. Permits therefore replace abatement costs and should be similarly treated for tax purposes in order to avoid distorting the abatement/permit tradeoff; although permits are capital assets, their cost should accordingly be deducted from gross income upon surrender.

Prior to actual use of the permit, the taxpayer cannot invoke a depreciation deduction because there is no ascertainable useful life over which it could be depreciated. Moreover, the permit does not experience gradual exhaustion, wear and tear, or obsolescence.

If the firm sells or exchanges an emission permit, the difference between the consideration paid to the firm (the amount realized) and its cost basis in the permit will be the taxable capital gain. The firm will recognize gain or loss in the year of the sale or exchange, unless a nonrecognition provision applies.

If the firm is a dealer in such permits, namely, it holds emission permits primarily for sale to customers in the ordinary course of trade or business of dealing in permits, any gain or loss realized from the sale or exchange will be ordinary income.

Penalties imposed for emitting without a permit, or beyond the level allowed by the permit, should not be deducted for income tax purposes. In the US, for example, section 162(f) of the Code provides that "no deduction shall be allowed under subsection (a) for any fine or similar penalty paid to the government for the violation of any law."

New Tax Challenges Created by Cap-and-Trade

Exacerbated Lock-in Effect If Permits Are Allocated Gratis

Income tax measures the taxpayer's potential to consume. A taxpayer's ability to consume is as much affected by a change in the value of her assets as by a change in the amount of cash she has. Nonetheless, changes in the value of an asset owned by the taxpayer are not taxed before a realization event takes place. Realization is a sale or other disposition of the property. The primary reason for the realization requirement is the difficulty in assessing the value of assets before they are actually sold. A secondary reason is liquidity problems that taxpayers may face if they are required to pay tax on an asset's appreciation prior to sale or disposition of the asset.

When a firm purchases a permit, either from the government in a primary auction or on the secondary market, it obtains a cost basis in the permit. It may use the permit in the current year to cover its emissions, deducting the cost upon surrender to the government, or it may bank the permit for use or sale in the future. If a firm decides to bank the permit, it must be expecting abatement costs (its own and others') to increase in the future at a rate that is higher than the yield it can earn on investment in other assets. The income tax's realization rule will, however, have a lock-in effect on the permit market. Firms will tend to defer permit sales that might otherwise be efficient in order to defer the tax on the accrued capital gain. This in turn will distort the permit/abatement tradeoff.

The lock-in effect will be especially significant when permits allocated gratis are not taxed upon receipt, as is the case under current US law (according to Rev. Rul. 92-16) and as is generally the case under the European Union Emissions Trading System (EU ETS). The firms have a zero tax-basis in their permits; hence the incentive to defer use or sale and the lock-in effect will be even greater. This increases demand for permits and distorts their market price upward, tending to result in inefficiently high levels of abatement.

Similarly, firms will be tax-induced to defer the use of their permits, that is, to continue banking them. Compared to other investment assets, investment in permits with zero-basis provides a tax-preferred return for the following reason. When a purchased asset is realized, the investor can deduct only the nominal (that is, historical) cost. This means that the amount invested in purchasing the asset is not even adjusted for inflation; hence inflationary gains are taxed, and the real value of the investment is decreased. No such out-of-pocket investment exists in case the permit was allocated gratis. This makes banking a permit that was received gratis a tax-preferred investment, distorting its price in equilibrium.

To reduce the distortion created by the lock-in effect, the government can auction the permits instead of allocating them gratis. Auction may be preferable on other efficiency grounds and on equity grounds as well. The cap creates scarcity and, by allocating the permits gratis, the government gives the scarcity rent to the firms, which is likely to have regressive effect to the extent that the rents are retained by firms rather than being passed on to consumers or labor. Moreover, the cap-and-trade system (even under a gratis allocation) raises the price of the underlying products by imposing a cost on products based on the emissions generated in their production, thereby lowering the real wage and distorting labor supply (as leisure cannot be taxed). This may create the same excess burden as a tax on labor income. If permits are auctioned, the revenue can potentially be used to reduce taxes on income and capital to correct for the inefficiency mentioned above, and/or to offset any regressive effects created if low-income people bear a larger share of the price increase. Of course, whether revenues are actually spent in these ways is politically contingent.

If, due to political constraints, the permits have to be allocated gratis, then they should be taxed on receipt. This will provide the government with revenue and will give the firms a tax basis equal to the fair market value of the permit on the date of receipt, thereby decreasing the lock-in effect to the same level as other assets.

Inventory Management Issues

A related issue is the inventory rule used in assessing taxes on stocks of assets. Firms which have purchased permits at different prices at different times have an incentive to surrender and deduct the costs of the more expensive permits while retaining those permits that were bought for low prices to sell in the long term in order to benefit from tax deferral, thereby exacerbating the lock-in effect. This additional effect can be prevented by requiring firms to manage their permits' use and sale on a first-in-first-out basis.

Alternatively, a firm's stock of permits could be valued and taxed annually on a mark-to-market basis. The values of all permits held by the firm are aggregated, based on their market values at the beginning and end of each year. The difference between the opening year balance and the end year balance is taxed. Sales and surrenders of permits throughout the year are deducted from the closing balance, and the proceeds from sales are included in taxable income. Eliminating the realization requirement in this way would eliminate the tax incentive for deferring use or sale of a permit and associated regulatory distortions.

The advantages of taxing capital assets on an accrual basis are well known, and the question of whether it is efficient to distinguish between traded assets, such as traded securities, and non-traded assets, whose value is difficult to ascertain, has been much debated. One could make a case for taxing tradable permits separately on an accrual basis.

Loss Limitation Rules

All countries with an income tax limit the deductibility of losses. They can only be used to offset gains (sometimes this requirement is eased by allowing some loss carry-forward to future tax years). Limited-loss deductibility introduces an asymmetry because gains are fully taxed but the taxpayer may not be able to deduct all losses. This asymmetry may exacerbate the lock-in effect as a result of permit price volatility. Firms may continue to hold permits in years in which they would otherwise sell them because if they did so they would incur losses that would not be fully tax deductible. This problem and the problems of market price volatility more generally (increasing uncertainty thereby resulting in sub-optimal production levels and in under-investment in innovation) can be addressed in the design of a cap-and-trade system by including safety valves to limit either excessively high or excessively low permit prices or both. Also, making the tax system symmetric with respect to gains and losses will reduce the cost to firms of permit price volatility and increase the efficiency of a cap-and-trade system. Symmetrical treatment could be limited to permits or applied to assets more generally. It is impossible to estimate, without empirical support, whether the inefficiencies are greater for permits than for any other assets, but there seems to be a consensus that a move to a more symmetric tax system would improve efficiency, and the treatment of permits could lead the way. Encouraging the development of markets for permit forwards, options, and swaps could assist in hedging the risk of price volatility, thereby increasing efficiency.

Transfer Pricing Problems

Countries tend to have quite different tax rates, and cap-and-trade creates a new possibility for tax arbitrage by multinational firms—purchasing and deducting permits in one country where the tax rate is high, though the actual production takes place in a second country where the tax rate is low. In order to deal with this problem, which will impair regulatory efficiency, countries must require multinational corporations to match the deductions of permits with the actual production whose emissions are being accounted for, and apply the same tax rate to both income and expense. This is already done by many countries in other contexts involving matching of income and expense items through transfer pricing rules. Cap-and-trade will add significantly to the importance of such rules and practices.

FURTHER READING

On the application of the problem of lock-in to the cap-and-trade system: Ethan Yale, "Taxing Cap and Trade Environmental Regulation," 37 *Journal of Legal Studies* 535 (2008).

Afterword

Reflections on a Path to Effective Climate Change Mitigation

Thomas Heller
Professor, Stanford University

Key Points

- There is a danger that in the international community's quest for a new climate agreement, we will lock things in too early around a weak arrangement, although the door is open for us to do much more.
- Two of the major challenges to reaching an international agreement are: uncertainly about the costs and effectiveness of mitigation efforts; and the conflict between developed countries that want to have global cap-and-trade and developing countries that do not.

There are many challenges along the path to a meaningful climate policy framework, but two stand out as particularly threatening. The first is uncertainty. More specifically, there is a serious risk that nations will not undertake meaningful action because of the persistence of uncertainty surrounding the relative cost and effectiveness of policies designed to mitigate climate change.

The second major challenge is the tension between the belief that a global cap-and-trade program is the best policy instrument to limit global greenhouse gas (GHG) emissions and the demand for fairness in allocating carbon caps among states, especially among developing nations.

Unfortunately, the debate has often seemed stuck on this tension, but recent actions by developing nations have pointed toward a different way forward. A growing chorus of voices is arguing that we need to quickly create a framework that will help encourage and finance bottom-up mitigation actions in developing countries even in the absence of caps. Despite the promise of this approach, it remains on the margins of the mainstream climate change debate.

In light of these developments, I am perhaps more afraid of a weak climate change agreement than no agreement at all. My fear is that a weak climate change agreement will result in complacency, and shut down efforts focused on building a framework to promote the changes that are already emerging out of the national policies of developing nations. This may be our greatest opportunity to mitigate global emissions reductions early, and we cannot afford to let it pass us by.

Uncertainty about Mitigation Benefits and Costs

Uncertainty can often have a paralyzing effect on both policymaking and investment. Societies and investors alike are averse to accepting policies with steep price tags when they are uncertain as to whether or not the benefits outweigh the costs. However, the risks of inaction are so great as to justify substantial investment in mitigation now. Recent reports (including the Stern Report) show that the costs of inaction outweigh, by a significant margin, the costs of action; that the current failure of markets to price carbon results in massive inefficiencies; and that the costs of postponing fixing the problem will only increase as time passes. However, the widespread resistance to climate change policies suggests that many politicians and voters do not believe in these conclusions or are afraid of the risk that costs will be much greater than predicted.

Obstacles to Global Application of Cap-and-Trade

A second fundamental challenge to our ability to limit global emissions in a timely fashion is the conflict between industrialized nations' drive towards a global cap-and-trade system and developing nations' resistance to national caps.

Industrialized nations have adopted or are adopting domestic cap-and-trade systems and have reached a consensus that a global cap-and-trade program would be the most efficient and effective means to address climate change. This consensus has emerged out of both scholarly literature and experience with actual policies, including failed attempts at imposing BTU taxes in the US and carbon taxes in Europe, as well as the success of SO_2 trading programs in both the US and Europe. This understanding has already been embodied in the cap-and-trade structure of the Kyoto Protocol's obligations for Annex I countries, as well as in the European Union Emissions Trading System (EU ETS). A major part of subsequent discussions has focused on increasing the participation in international cap-and-trade until it encompasses all nations, or at least all major emitters.

However, as is often the case in international negotiations, there is a countervailing principle—common but differentiated responsibility. This principle centers on the recognition that although all nations bear some responsibility to address global environmental problems, the scope of their obligations vary according to a wide variety of legitimate concerns, all of which push against an easy or straightforward application of global cap-and-trade. Developing countries strongly resist caps. They point to the developed countries' historical responsibility for the greater part of the emissions that are causing warming today. Moreover, they are deeply concerned that the adoption of national caps will hem in their future economic growth in a way that is extremely constrictive and unfair.

Because of the resulting impasse, negotiations have been stuck in a bind for some time. On the one hand, we acknowledge that cap-and-trade is the most efficient solution. On the other hand, we are unable to resolve the distributional problems necessary to implement it on a global scale.

In the shadow of this debate, separate discussions have grown around alternative means of financing mitigation actions in developing countries, as seen in many chapters in this book. However, these alternatives have not been fully embraced by either industrialized or developing nations. Industrialized nations tend to view climate finance alternatives to global cap-and-trade as partial solutions at best, a distraction from the larger push toward cap-and-trade. Developing nations, on the other hand, tend to view many of these proposed mechanisms suspiciously, in part because they fear that they are a hook to draw them into binding caps.

Consequences of the Conflict

The consequence of this conflict and the resulting turn toward smaller, narrower discussions has been a balkanization of climate change negotiations. At present there are many special negotiations and working groups focused on specific issues, such as technology transfers, flexibility mechanisms, comparability, carbon finance, and deforestation. This process has both advantages and disadvantages. The advantages are that it helps policymakers refine specific policies, begin to implement them, and gain a greater understanding of the difficulties they and other similar policies will present. For example, discussions about forestry have resulted in the development of a wide portfolio of proposed programs for reducing emissions from deforestation and forest degradation (REDD), a more refined understanding of how to create an international framework for REDD crediting, as well as an understanding of the broader challenges in implementing REDD and complex problems presented by sectoral caps or crediting baselines generally.

However, the disadvantages to this micro-policy approach are that parties begin to excessively focus on small victories, reduce their expectations, and lose sight of the main goal—creating a framework that will facilitate and encourage global mitigation actions on the scale necessary to avert catastrophic warming. In other words, we may end up with a lot of small projects that yield only small benefits and overall are not particularly efficient or effective, at least when viewed from a global perspective.

An Alternative Path Forward

Accordingly the view from the top is bleak. However, an entirely different picture emerges when one begins to look at national-level actions that are occurring across a wide variety of nations. Increasingly, both developed and developing countries are beginning to view high-carbon economic growth as an oxymoron, because of fears that the negative consequences of high-carbon growth will ultimately undercut the gains reaped from such growth. As a result, we are beginning to see changes in developing countries' national policies that are consistent with the idea of low-carbon growth. Even more promising, these efforts become part of international negotiations. This position has perhaps been stated most clearly by South Africa, which said: we will do what is in our self-interest; we will do something more than that because we are part of the global community, and there are things we will do still further with support from those who are better positioned to help us. To realize this, policymakers must answer the following questions: how do we increase mitigation efforts in developing nations in the absence of binding targets, and how do we best structure and scale up financial and technical assistance from developed to developing nations in the absence of national caps? It is to be hoped that this book provides useful answers to these questions.

Abbreviations

AAU	Assigned Amount Unit under the Kyoto Protocol
ACES	American Clean Energy and Security Act
AFB	Adaptation Fund Board
AFOLU	agriculture, forestry, and land use
BAU	business as usual
BCA	border climate adjustment
BTA	border tax adjustment
CC	climate change
CCS	carbon capture and storage
CDM	Clean Development Mechanism
CER	Certified Emissions Reduction under the CDM
CFL	carbon footprint labeling
CFU	World Bank Carbon Finance Unit
CITES	Convention on International Trade in Endangered Species
CLEAR	Carbon Limits + Early Actions = Rewards
CO_2e	carbon dioxide equivalent
COP	Conference of the Parties to the UNFCCC
COP 13	Bali Conference of the Parties
CTF	World Bank Clean Technology Fund
DDA	Doha Development Agenda
DEFRA	UK Department for Environment, Food, and Rural Affairs
EDF	Environmental Defense Fund
EGS	environmental goods and services
EIO	environmental input-output
EPA	US Environmental Protection Agency
ETS	emissions trading system
EU	European Union
EU ETS	European Union Emissions Trading System
GATS	WTO General Agreement on Trade in Services

GATT	WTO General Agreement on Tariffs and Trade
GDP	gross domestic product
GEF	Global Environment Facility
GHG	greenhouse gas
GNP	gross national product
IBRD	International Bank for Reconstruction and Development
IEA	International Energy Agency
IFC	International Finance Corporation
IMF	International Monetary Fund
INCR	Investor Network on Climate Risk
IPAM	Instituto de Pesquisa Ambiental da Amazônia, or Amazon Institute for Environmental Research
IPCC	Intergovernmental Panel on Climate Change
IPRs	intellectual property rights
ISO	International Organization for Standardization
JI	Joint Implementation
KP	Kyoto Protocol
LDCs	least developed countries
LEED	Leadership in Energy and Environmental Design
LULUCF	land use, land use change, and forestry
MARPOL	International Convention for the Prevention of Pollution from Ships
MEA	multilateral environmental agreement
MFN	most favored nation
MIGA	Multilateral Investment Guarantee Agency
MOP	Meeting of the Parties to the Kyoto Protocol
MRV	monitoring, verification, and reporting
NAI	Non–Annex I
NAMA	nationally appropriate mitigation action
NC	National Communication
NGO	non-governmental organization
ODA	official development assistance
OECD	Organisation for Economic Co-operation and Development
OPIC	Overseas Private Investment Corporation
PAS	Publicly Available Standard
ppmv	parts per million by volume
R&D	research and development
REC	renewable energy certificate or renewable energy credits
REDD	reducing emissions from deforestation and forest degradation

RFM	Reformed Financial Mechanism
RPS	renewable power standards
RTA	regional trade agreement
SCM	WTO Subsidies and Countervailing Measures Agreement
SD-PAMs	sustainable development policies and measures
SEC	US Securities and Exchange Commission
SNLT	sectoral no-lose target
SPS	WTO Agreement on the Application of Sanitary and Phytosanitary Measures
TBT	WTO Agreement on Technical Barriers to Trade
TPRM	WTO Trade Policy Review Mechanism
TRIPS	WTO Agreement on Trade-Related Aspects of Intellectual Property Rights
TRIMS	WTO Agreement on Trade-Related Investment Measures
UNCED	UN Conference on Environment and Development
UNDP	United Nations Development Programme
UNFCCC	United Nations Framework Convention on Climate Change
USAID	US Agency for International Development
WBCSD	World Business Council for Sustainable Development
WTO	World Trade Organization
WWF	World Wide Fund for Nature

Index